Isaac Chemistry

Developing mastery of

Essential pre-university physical chemistry

D.I. FOLLOWS
Head of Chemistry
Winchester College

Periphyseos Press
Cambridge, UK.

Co-published in Cambridge, United Kingdom, by
Periphyseos Press and Cambridge University Press.

www.periphyseos.org.uk and www.cambridge.org

First published 2016. First, second & third reprints, 2017.
Fourth reprint 2018. Fifth reprint 2020.
Co-published adaptation, 2020

Printed and bound in the UK by Short Run Press Limited, Exeter.

Typeset in LaTeX

A catalogue record for this publication is available from the British Library

ISBN 978-1-8382160-3-0 Paperback

Use this book in parallel with the electronic version at
isaacchemistry.org
Marking of answers and compilation of results is free on Isaac Chemistry.
Register as a student or as a teacher to gain full functionality and support.

Cover images: Atomic orbitals (created with WolframMathematica®.)

used with kind permission of M. J. Rutter.
'The Periodic Table of the Elements' reproduced with kind permission of OCR

Contents

Notes for the Student and the Teacher **iv**

Using Isaac Chemistry with this book; Significant figures **v**

Mapping of Book Sections to Exam Specifications **vi**

Table of Constants **viii**

A Formulae & Equations **1**
- A1 Empirical formulae 1
- A2 A_r & M_r and molecular formula 2

B Amount of Substance **3**
- B1 Standard form 3
- B2 Unit conversions 4
- B3 Gases 5
- B4 Solids 6
- B5 Solutions 6
- B6 Reactions 7
- B7 Titration 8
- B8 Parts per million 11

C Gas Laws **12**
- C1 Ideal gases 12
- C2 Density 15
- C3 Non-ideal gases 16

D Atomic Structure **17**
- D1 Atomic structure 17
- D2 Atomic orbitals 18

D3 Atomic and ionic radii and ionization energy 19
D4 Isotopes 21

E Electronic Spectroscopy **23**

F Enthalpy Changes **25**
F1 Calorimetry 25
F2 Bond enthalpies 28
F3 Formation and combustion enthalpies 29
F4 Born-Haber cycles 31

G Entropy **35**
G1 Absolute entropy 35
G2 Entropy changes 36

H Free Energy **37**
H1 Entropy change of the surroundings 37
H2 Free energy changes 39
H3 Free energy cycles 40
H4 Thermodynamic stability 42

I Equilibrium **43**
I1 Equilibrium constant, K_p 43
I2 Equilibrium constant, K_c 45
I3 Solubility product 49
I4 Partition 51
I5 $RT \ln K$ 53

J Acids & Bases **55**
J1 Brønsted-Lowry & Lewis 55
J2 pH & K_w 57
J3 K_a & pK_a 59
J4 K_b & pK_b 60
J5 Buffers 60

K Redox **62**
K1 Oxidation number 62
K2 Half-equations 63
K3 Balancing redox equations 64
K4 Disproportionation 66

L Electrochemistry **67**
L1 Electrode potential & cell potential 67
L2 Free energy & K_c 68
L3 Spontaneous redox reactions 69

M Rate Laws, Graphs & Half-life **70**
M1 Rate laws 70
M2 Half-life 75
M3 The Arrhenius model 78
M4 Catalysis 81
M5 Michaelis-Menten kinetics 83

Notes for the Student and the Teacher

- "Developing mastery of essential pre-university physical chemistry" is associated with the **O**pen **P**latform for **A**ctive **L**earning : `isaacchemistry.org`
- The problems in this book are designed, graded and arranged by sections for understanding and mastery of the concepts and problem solving required at A-level chemistry and for equivalent exams.
- A mapping of each section of the book on to the relevant parts of all major UK chemistry exam specification is given on pages vi and vii. [With thanks to Dr David Paterson.] On-line, the mapping at `isaacchemistry.org/pages/syll_map_chem` allows 2-click setting of a chosen section as homework. See also below for teacher functionality.
- Each question can be answered on the above OPAL where there is immediate marking and some feedback. Practice to achieve *mastery* will give fluency and confidence in the essential skills required for strong A-level results and in the transition to university. Mastery level, that is achieving at least 80% of questions correct, is indicated in the marginal square boxes .

$^{n}/_{m}$

- Students should register (free) on Isaac. Go to `isaacchemistry.org` – click the "log in" tab and then "sign up" if not already registered. Physics registration on `isaacphysics.org` carries over for chemistry, and *vice versa*. A benefit of registration is that all activity is recorded for the student. Should questions be later set as homework, they will automatically be noted as done. All activity is confidential to the student, except parts that the student decides to share with a teacher (e.g. set home work).
- Teachers who register can have their Isaac accounts converted to teacher accounts (see at the foot of the user profile page). They can then set home work, and have it automatically marked with results returned to them: See "For Teachers" at `isaacchemistry.org` or go directly to `isaacchemistry.org/teacher_features`. Chemistry and physics functionalities are the same and are activated by registration on either the Physics or Chemistry sites.
- Visit `isaacchemistry.org` for physical chemistry problems in addition to these chemistry book problems (currently at "Further problem solving"). Be sure to select the level required, or otherwise get questions from all levels. These questions are all chemical thermodynamics for pre-AS, AS, A2, & the transition to university, with reaction kinetics soon to follow.
- See the physics companion book, also at `www.isaacbooks.org/`. It too is available for £1 per copy. Problems from the physics book can be answered on-line at `isaacphysics.org/physics_skills_19`.
- Maths problems to give the fluency required at chemistry and physics A-level, and equivalents, are available on `isaacphysics.org`.

Using Isaac Chemistry with this book

Isaac Chemistry offers on-line versions of each sheet at `isaacchemistry.org/book16` where a student can enter answers. This on-line tool will mark answers, giving immediate feedback to a student who, if registered on `isaacchemistry.org`, can have their progress stored and even retrieved for their CV! Teachers can set a sheet for class homework as the appropriate theme is being taught, and again for pre-exam revision. Isaac Chemistry can return the fully assembled and analysed marks to the teacher, if registered for this free service. Isaac Chemistry zealously follows the significant figures (sf) rules below and warns if your answer has a sf problem.

Uncertainty and Significant Figures

In chemistry, numbers represent values that have uncertainty and this is indicated by the number of significant figures in an answer.

Significant figures

When there is a decimal point (dp), all digits are significant, except leading (leftmost) zeros: 2.00 (3 sf); 0.020 (2 sf); 200.1 (4 sf); 200.010 (6 sf)

Numbers without a dp can have an *absolute accuracy*: 4 people; 3 electrons.

Some numbers can be ambiguous: 200 could be 1, 2 or 3 sf (see below). Assume such numbers have the same number of sf as other numbers in the question.

Combining quantities

Multiplying or dividing numbers gives a result with a number of sf equal to that of the number with the smallest number of sf:

$x = 2.31$, $y = 4.921$ gives $xy = 11.4$ (3 sf, the same as x).

An absolutely accurate number multiplied in does not influence the above.

Standard form

On-line, and sometimes in texts, one uses a letter 'x' in place of a times sign and ^ denotes "to the power of":

1800000 could be 1.80x10^6 (3 sf) and 0.0000155 is 1.55x10^-5

(standardly, 1.80×10^6 and 1.55×10^{-5})

The letter 'e' can denote "times 10 to the power of": 1.80e6 and 1.55e-5.

Significant figures in standard form

Standard form eliminates ambiguity: In $n.nnn \times 10^n$, the numbers before and after the decimal point are significant:

$191 = 1.91 \times 10^2$ (3 sf); 191 is $190 = 1.9 \times 10^2$ (2 sf); 191 is $200 = 2 \times 10^2$ (1 sf).

Answers to questions

In this book and on-line, give the appropriate number of sf: For example, when the least accurate data in a question is given to 3 significant figures, then the answer should be given to three significant figures; see above. Too many sf are meaningless; giving too few discards information. Exam boards also require consistency in sf.

		OCR A (H032/H432)	OCR B (H033/H433)	AQA (7404/7405)	Edexcel (8CH0/9CH0)	Eduqas (A410QS)	CIE Pre-U (9791)	IB Chemistry (Higher)
A	**Formulae & Equations**							
A1	Empirical formulae	2.1.3	EL(b)	3.1.2.4	Topic 5	C1.3	A4.1	1.2
A2	A_r & M_r and molecular formula	2.1.3	EL(a)	3.1.1.2 / 3.1.2.1	Topic 1	C1.3	A4.1	1.2
B	**Amount of Substance**							
B1	Standard form	throughout	throughout	throughout	throughout	throughout	throughout	throughout
B2	Unit conversions	throughout	throughout	throughout	throughout	throughout	throughout	throughout
B3	Gases	2.1.3	DF(a)	3.1.2.2	Topic 5	C1.3	A1.1	1.2
B4	Solids	2.1.3	EL(b)	3.1.2.2	Topic 5	C1.3	A1.1	1.2
B5	Solutions	2.1.3	DF(a)	3.1.2.2	Topic 5	C1.3	A4.1	1.3
B6	Reactions	2.1.3	EL(b)	3.1.2.5	Topic 5	C1.3	A1.1	1.3
B7	Titration	2.1.4	EL(c)	3.1.12.5	Topic 5	C1.3	A4.1	1.3
B8	Parts per million	2.1.3	OZ(i)					1.3
C	**Gas Laws**							
C1	Ideal gases	2.1.3	DF(a)	3.1.2.3	Topic 5	C1.3	B1.7	1.3
C2	Density							1.3
C3	Non-ideal gases							1.3
D	**Atomic Structure**							
D1	Atomic structure	2.2.1	EL(f), DM(h)	3.1.1.3	Topic 1	C1.2	A1.2	2.2
D2	Atomic orbitals	2.2.1	EL(e)	3.1.1.3	Topic 1	C1.2	A1.2	2.2
D3	Atomic and ionic radii and ionization energy	3.1.1	EL(q)	3.1.1.3	Topic 1	C1.2	A1.2	3.2
D4	Isotopes	2.1.1	EL(h)	3.1.1.2	Topic 1	C1.3	A1.1	2.1, C.3, D.8
E	**Electronic Spectroscopy**							
	Electronic Spectroscopy		EL(w)	3.2.5.4	Topic 1	C1.2	A4.3	2.2
F	**Enthalpy Changes**							
F1	Calorimetry	3.2.1	DF(f)	3.1.4.2	Topic 8	C2.2	A1.4	5.1
F2	Bond enthalpies	3.2.1	DF(e)	3.1.4.4	Topic 8	C2.2	A1.4	5.3
F3	Formation and combustion enthalpies	3.2.1	DF(g)	3.1.4.3	Topic 8	C2.2	A1.4	5.1
F4	Born-Haber cycles	5.2.1	(O(b))	3.1.8.1	Topic 13A	PI4.1	A1.4	15.1
G	**Entropy**							
G1	Absolute entropy	5.2.2	O(g)	3.1.8.2	Topic 13B	PI4.2	B1.5	15.2
G2	Entropy changes	5.2.2	O(g)	3.1.8.2	Topic 13B	PI4.2	B1.5	15.2

		OCR A (H032/H432)	OCR B (H033/H433)	AQA (7404/7405)	Edexcel (8CH0/9CH0)	Eduqas (A410QS)	CIE Pre-U (9791)	IB Chemistry (Higher)
H	**Free Energy**							
H1	Entropy change of the surroundings	5.2.2	O(f)	3.1.8.2	Topic 13B		B1.5	15.2
H2	Free energy changes	5.2.2		3.1.8.2	Topic 13B	PI4.2	B1.5	15.2
H3	Free energy cycles						B1.5	
I	**Equilibrium**							
I1	Equilibrium constant, K_p	5.1.2		3.1.10	Topic 11	PI5	B1.6	
I2	Equilibrium constant, K_c	3.2.3, 5.1.2	ES(p), CI(h)	3.1.6.2	Topic 11	C2.1, PI5.1	B1.6	7.1, 17.1
I3	Solubility product		O(h)				B1.6	A.10
I4	Partition							D.3
I5	$RT \ln K$				Topic 13B		B1.5	15.2, 17.1
J	**Acids & Bases**							
J1	Brønsted-Lowry & Lewis	5.1.3	O(i)	3.1.12.1	Topic 12	PI5.2	B1.6	8.1, 18.1
J2	pH & K_w	5.1.3	O(l)	3.1.12.2 / 3.1.12.3	Topic 12	PI5.2	B1.6	8.3
J3	K_a & pK_a	5.1.3	O(l)	3.1.12.4	Topic 12	PI5.2	B1.6	18.2
J4	K_b & pK_b						(B1.6)	18.2
J5	Buffers	5.1.3	O(m)	3.1.12.6	Topic 12	PI5.2	B1.6	18.2
K	**Redox**							
K1	Oxidation number	2.1.5	ES(e), DM(c)	3.1.7	Topic 3	C1.1, C1.6	B1.6	9.1
K2	Half-equations	5.2.3	ES(d), DM(c)	3.1.7	Topic 3	PI1.1	B1.6	9.1, 19.1
K3	Balancing redox equations	5.2.3	DM(c)	3.1.7	Topic 3	PI1.2	A2.1, B1.6	19.1, A.2
K4	Disproportionation	3.1.3, 5.3.1		3.1.7	Topic 4B	PI2.1	A2.2	
L	**Electrochemistry**							
L1	Electrode potential & cell potential	5.2.3	DM(f)	3.1.11.1	Topic 14	PI1.1	B1.6	19.1
L2	Free energy & K_c	(5.2.3)	(DM(f))		(Topic 14)	(PI1.1)	B1.6	19.1
M	**Rate Laws, Graphs & Half-life**							
M1	Rate laws	5.1.1	CI(a)	3.1.9.1	Topic 16	PI3	B1.7	16.1
M2	Half-life	5.1.1	CI(b)		Topic 16		B1.7	
M3	The Arrhenius model	5.1.1	CI(d)	3.1.9.1	Topic 16	PI3	B1.7	16.2
M4	Catalysis	various incl. 3.2.2, 3.2.3, 5.1.2	various in DF, OZ, CI, CM	various incl. 3.1.5, .6, .10	Topic 9, 16	various incl. C2.3, PI2.2, PI3	A1.4, B1.7	various incl. 6.1, 7.1, 16.1, A.3
M5	Michaelis-Menten kinetics		(PL(f))				(B1.7)	B.7

Table of Constants

Quantity	**Magnitude**	**Unit**
Gas constant (R)	8.31	$J\ mol^{-1}\ K^{-1}$
Electron volt (eV)	1.60×10^{-19}	J
Avogadro's number (N_A)	6.02×10^{23}	–
Boltzmann's constant	1.38×10^{-23}	$J\,K^{-1}$
Speed of light (*in vacuo*)	3.00×10^{8}	$m\,s^{-1}$
Atomic mass unit (u)	1.66054×10^{-27}	kg
Planck's constant (h)	6.63×10^{-34}	J s
Charge on electron	1.60×10^{-19}	C
0 °C	273.15	K
Specific heat capacity of water	4180	$J\,kg^{-1}\,K^{-1}$
Density of water at RTP	1.00	$g\,cm^{-3}$
Faraday constant	96485	$C\,mol^{-1}$
Volume of gas at RTP	24	$dm^3\,mol^{-1}$
Room Temperature	298	K
Room Pressure = 1 Atm.	1.013×10^{5}	Pa

Chapter A

Formulae & Equations

The boxed fraction shows how many questions need to be answered correctly to achieve mastery.

A1 Empirical formulae

10/12

A1.1 Find the empirical formulae for the ten compounds (a)–(j) from the data given. No compound contains more than 15 atoms in total in its formula. Show all your working in neat, clearly-presented answers. All compositions are by mass.

a) 35.0% Nitrogen, 5.0% Hydrogen, 60.0% Oxygen

b) 90.7% Lead, 9.3% Oxygen

c) 26.6% Potassium, 35.3% Chromium, 38.1% Oxygen

d) 40.3% Potassium, 26.8% Chromium, 32.9% Oxygen

e) 29.4% Vanadium, 9.2% Oxygen, 61.4% Chlorine

f) 81.8% Carbon, 18.2% Hydrogen

g) 38.7% Carbon, 9.7% Hydrogen, 51.6% Oxygen

h) 77.4% Carbon, 7.5% Hydrogen, 15.1% Nitrogen

i) 25.9% Nitrogen, 74.1% Oxygen

j) 29.7% Carbon, 5.8% Hydrogen, 26.5% Sulphur, 11.6% Nitrogen, 26.4% Oxygen

Element	Atomic mass	Element	Atomic mass
Hydrogen	1.0	Chlorine	35.5
Carbon	12.0	Potassium	39.1
Nitrogen	14.0	Vanadium	50.9
Oxygen	16.0	Chromium	52.0
Sulphur	32.1	Lead	207.2

A1.2 Complete combustion of 6.4 g of compound K produced 8.8 g of carbon dioxide and 7.2 g of water. Calculate the empirical formula of K.

A1.3 Complete combustion of 1.8 g of compound L produced 2.64 g of carbon dioxide, 1.08 g of water and 1.92 g of sulfur dioxide. Calculate the empirical formula of L.

8/10

A2 A_r & M_r and molecular formula

Assume that the mass of an isotope in amu *to* 3 *s.f. is equal to its mass number.*

A2.1 Which isotope is used as the standard against which relative atomic masses are calculated?

A2.2 Fluorine only occurs naturally in one isotope, ^{19}F, and has a relative atomic mass of 19.0 amu. Calculate the mass of a fluorine atom in kg.

A2.3 Magnesium has the following natural isotopes: ^{24}Mg 78.6%; ^{25}Mg 10.1%; ^{26}Mg 11.3%. Calculate the relative atomic mass of magnesium.

A2.4 The relative atomic mass of boron is 10.8 amu. Boron exists in two isotopes, ^{10}B and ^{11}B. Calculate the percentage abundance of ^{10}B.

A2.5 Complete the table below:

ELEMENT	A_r	Isotope 1	Isotope 2	Isotope 3	Isotope 4
Bromine	(a)	^{79}Br 50.5%	^{81}Br 49.5%	n/a	n/a
Silver	107.9	(b) ^{107}Ag **?**%	(c) ^{109}Ag **?**%	n/a	n/a
Cerium	140.2	^{136}Ce 0.2%	^{138}Ce 0.2%	^{140}Ce 88.5%	(d) $^{\mathbf{?}}Ce$ 11.1%

A2.6 The relative molecular mass of compound M is 135 amu. M contains 3.7% hydrogen, 44.4% carbon and 51.9% nitrogen by mass. Find the molecular formula of M.

A2.7 Complete combustion of compound N occurs in a stoichiometric ratio of 1 : 6 with oxygen gas. Complete combustion of 4.2 g of compound N produces 13.2 g of carbon dioxide and 5.4 g of water. Find the molecular formula of N.

Chapter B

Amount of Substance

B1 Standard form

$^{12}/_{14}$

B1.1 Complete the following calculations, giving the answers in standard form to the appropriate number of significant figures. For guidance on this and on how to enter your answers on `isaacchemistry.org`, consult page ii of this book.

a) 120×70

b) $5600 + 800 + 12 + 1100 + 320$

c) $\dfrac{95.0}{19\,000}$

d) $12000 + 84000 + (3.00 \times 10^{3}) + 29000$

e) $(4.0 \times 10^{2}) \times 100 \times 300$

f) $\dfrac{1.6 \times 10^{-8}}{6.4 \times 10^{-3}}$

g) $\dfrac{3.00 \times 10^{8}}{5.2 \times 10^{-7}}$

h) $2.12 \times 10^{12} \times 5.4 \times 10^{6}$

i) $1.4 \times 10^{-10} \times 1.4 \times 10^{-10} \times 2.2 \times 10^{-10}$

j) $1.6 \times 10^{-19} \times 6.0 \times 10^{23}$

k) $\dfrac{1.3 \times 10^{17}}{3.0 \times 10^{8}}$

l) $(1.4 \times 10^{-6})^{3}$

m) $\sqrt{2.5 \times 10^{14}}$

n) $\dfrac{2.0 \times 10^{4} \times 1.2 \times 10^{4}}{(3.2 \times 10^{6})^{2}}$

o) $\dfrac{1.1 \times 10^{-5} \times (-2) \times 3}{(9.6 \times 10^{-11} + 1.2 \times 10^{-10})}$

27/33

B2 Unit conversions

Use standard form where answers are outside the range 0.01 to 1000 units.

B2.1 Convert the following volumes into dm^3

a) $0.86\ m^3$
b) $200\ cm^3$
c) 45 ml
d) $120\ m^3$
e) $0.064\ nm^3$
f) 70 cl
g) $1.6\ mm^3$
h) 1100 cc
i) $2.2\ km^3$
j) $42.5\ Å^3$

B2.2 Converts the following masses into g:

a) 16.0 kg
b) 120 mg
c) 0.004 kg
d) 12 tonne
e) $54\ \mu g$

B2.3 Convert the following into standard SI units:

a) $68\ km\,h^{-1}$
b) 500 g
c) $24\ dm^3$
d) 20 mbar
e) $-77\ °C$
f) 5.0 h
g) 740 nm
h) $72\ mN\,cm^{-1}$
i) 1014 mbar
j) $13.8\ g\,cm^{-3}$

B2.4 Give the results of the following calculations in standard SI units:

a) Density = $250\ g / 400\ cm^3$
b) Speed = 96 km / 80 min
c) Concentration = $2.50\ mmol / 40.0\ cm^3$ (use $mol\,dm^{-3}$)
d) Momentum = $4.0 \times 10^{-23}\ g \times 900\ m\,s^{-1}$
e) Pressure = $590\ fN / 10\ nm^2$
f) Volume = $240\ pm \times 240\ pm \times 320\ pm$
g) Amount = $2.0\ \mu mol\,dm^{-3} \times 75\ \mu m^3$
h) Energy = $3.2 \times 10^{-19}\ C \times 2.4\ kV$

B3 Gases

16/19

RTP = room temperature and pressure.
Any gas occupies 24 dm^3 per mole at RTP.
Avogadro's number, $N_A = 6.02 \times 10^{23}$.

B3.1 Calculate the volume occupied by:

a) 4.0 moles of gas at RTP

b) 0.030 moles of gas at RTP

c) 5.0×10^{18} atoms of helium gas at RTP

d) 1.2×10^{24} molecules of ozone at RTP

e) 8.0 g of O_2 at RTP

f) 1.1 kg of carbon dioxide at RTP

B3.2 Calculate the amount of gas (at RTP) in:

a) 4.8 dm^3

b) 12 m^3

c) 400 cm^3

d) 18 ml

B3.3 Calculate the number of molecules of gas (at RTP) in:

a) 36 dm^3

b) 300 cm^3

B3.4 Calculate the number of atoms (at RTP) in:

a) 60 cm^3 of argon

b) 1.2 dm^3 of N_2

c) 8.0 m^3 of carbon dioxide

d) 420 cm^3 of ethene

B3.5 Calculate the the mass of:

a) 1.0 m^3 of neon at RTP

b) 20 cm^3 of $(CH_3)_2O$ at RTP

c) 420 cm^3 of ammonia at RTP

18/22

B4 Solids

B4.1 Find the molar masses in amu of the following compounds:

a) $CaCO_3$
b) Na_2CO_3
c) NaOH
d) HCl
e) H_2SO_4
f) $FeSO_4$
g) $KMnO_4$
h) $Fe_2O_3{\cdot}5\,H_2O$ [1]
i) Calcium hydroxide
j) Butane

B4.2 Calculate the mass of:

a) 0.25 moles of H_2O_2(l)
b) 6.0 moles of C_2H_6(g)
c) 0.40 moles of H_2O(l)
d) 20.0 moles of Sr(s)
e) 1.20 moles of aluminium oxide
f) 7.4 moles of ammonium sulfate

B4.3 Calculate the amount of:

a) 1.001 g of $CaCO_3$(s)
b) 197 kg of Au(s)
c) 1.4 g of CO(g)
d) 2.006 kg of Hg(l)
e) 11.1 g of lithium carbonate
f) 10.0 mg of lead(II) iodide

10/12

B5 Solutions

B5.1 Calculate the concentration in mol dm^{-3} of the following solutions:

a) 0.40 g NaOH in 100 ml water
b) 7.3 g HCl in 1000 ml water
c) 2.5 g H_2SO_4 in 50 ml water
d) 15 g $FeSO_4$ in 500 ml water
e) 0.16 g $KMnO_4$ in 200 ml water

B5.2 Calculate the mass of solute in each of the following:

a) 500 ml of 0.010 mol dm^{-3} NaOH
b) 150 ml of 4.0 mol dm^{-3} HCl
c) 1.00 ml of 10.0 mol dm^{-3} H_2SO_4

[1] The 5 means that the formula includes 5 of what follows, i.e. water, so total mass is for $Fe_2O_3 + 5 \times H_2O$

d) 25.0 ml of 0.50 mol dm^{-3} $FeSO_4$

e) 21.8 ml of 0.0050 mol dm^{-3} $KMnO_4$

f) 2.0 dm^3 of 0.10 mol dm^{-3} NaCl

g) 100 ml of limewater with a concentration of 0.00020 mol dm^{-3}

B6 Reactions

B6.1 Calculate the amount of oxygen needed, and amount of carbon dioxide produced, in each of the following cases:

a) $C_3H_8 + 5\,O_2 \longrightarrow 3\,CO_2 + 4\,H_2O$, using 1.0 mole of C_3H_8

b) $C_2H_6O + 3\,O_2 \longrightarrow 2\,CO_2 + 3\,H_2O$, using 0.2 moles of C_2H_6O

c) $2\,CO + O_2 \longrightarrow 2\,CO_2$, using 4.0 moles of CO

d) $C_6H_{12}O_6 + 6\,O_2 \longrightarrow 6\,CO_2 + 6\,H_2O$, using 0.040 moles of $C_6H_{12}O_6$

e) $C_2H_4O_2 + 2\,O_2 \longrightarrow 2\,CO_2 + 2\,H_2O$, using 0.10 moles of $C_2H_4O_2$

B6.2 Calculate the amount of water produced by complete combustion of the following in oxygen (you will need to write a balanced equation each time):

a) 1.0 mole of pentane, C_5H_{12}

b) 2.5 moles of heptane, C_7H_{16}

c) 200 moles of hydrogen, H_2

d) 4.0 moles of butane

e) 0.0030 moles of methane

B6.3 Write the equation for each reaction and hence calculate the amount of acid required for complete reaction in each of the following cases:

a) 0.10 mol NaOH reacting with H_2SO_4

b) HCl reacting with 20 g of $CaCO_3$

c) 24 g CuO reacting with HNO_3

d) 5.6 g Fe reacting with HCl

e) 14.8 g of calcium hydroxide reacting with H_2SO_4

f) 10 g of magnesium oxide reacting with nitric acid

B6.4 Calculate the volume of 0.50 mol dm^{-3} H_2SO_4 required to neutralize each of the following:

a) 25.0 cm^3 of 1.0 mol dm^{-3} NaOH

b) 3.0 g $CaCO_3$

c) 1.25 g $ZnCO_3$

d) 4.03 kg MgO

e) 100 cm^3 of 0.2 mol dm^{-3} $NH_3(aq)$

20/24

B7 Titration

B7.1 Nitric acid of unknown concentration was added to a burette. 25.00 cm^3 of potassium hydroxide solution at a concentration of 0.100 mol dm^{-3} was transferred to a 250 cm^3 conical flask using a volumetric pipette. A few drops of methyl orange indicator were added to the flask.

a) Give the colour of the indicator in the alkaline solution in the flask.

The nitric acid was added to the flask a little at a time until the resulting solution went pink. The whole process was repeated until concordant titres (within 0.10 cm^3) were obtained.

b) The concentration of nitric acid was found to be 0.092 mol dm^{-3}. Calculate the titre obtained, in cm^3.

B7.2 In an analysis, 2.50 g of an unknown carbonate were dissolved in 100 cm^3 of 1.00 mol dm^{-3} hydrochloric acid (an excess). The resulting solution was made up to 250 cm^3 in a volumetric flask. 25.00 cm^3 aliquots of this solution were titrated against 0.250 mol dm^{-3} sodium hydroxide. Some of the results are shown below. Fill in the gaps in the table overleaf, and then calculate the quantities below to identify the cation.

TITRATION	Initial burette reading / cm^3	Final burette reading / cm^3	Titre / cm^3
ROUGH	0.60	25.10	(a)
1	0.15	(b)	24.10
2	(c)	25.25	24.45
3	1.35	25.45	(d)

e) Calculate the average concordant titre.

f) Calculate the amount of sodium hydroxide in that volume.

g) Calculate the amount of hydrochloric acid in each aliquot.

h) Calculate the initial amount of hydrochloric acid added to the carbonate.

i) Calculate the amount of hydrochloric acid remaining after reaction.

j) Calculate the amount of hydrochloric acid used in reaction with the carbonate.

k) Calculate the amount of carbonate in 2.50 g.

l) Calculate the molar mass of the carbonate.

m) Identify the cation in the carbonate.

B7.3 All of the ozone in 5.00 m^3 of air was reacted with 250 cm^3 of potassium iodide solution:

$$O_3 + 2\,I^- + 2\,H^+ \longrightarrow I_2 + O_2 + H_2O.$$

The liberated iodine was titrated against a standard solution of sodium thiosulfate with a concentration of 0.0400 mol dm^{-3}. 25.0 cm^3 of the iodine solution was used in each titration. The results of the titration are shown in the table overleaf. Fill in the remaning titres, and then answer the questions which follow.

TITRATION	Initial burette reading / cm^3	Final burette reading / cm^3	Titre / cm^3
ROUGH	0.10	25.40	25.30
1	0.80	26.10	(a)
2	1.20	26.20	(b)
3	1.00	25.90	(c)

d) Calculate the concentration of the iodine solution.

e) Calculate the amount of ozone in the 5.00 m^3 of air.

f) Name the piece of apparatus that should be used to transfer the iodine solution into a conical flask, ready for titration.

g) Name a suitable indicator for this titration, and give its colour change at the end point.

B7.4 Three students each prepare a standard solution by dissolving 10.6 g of solid from different bottles labelled 'sodium carbonate' in exactly 1 dm^3 of water. They use this standard solution in a titration to determine the exact concentration of a solution of sulphuric acid at approximately 0.1 $mol\,dm^{-3}$. They each use a pipette to measure out exactly 25.00 cm^3 of the standard solution into a conical flask; they use the same indicator and carry out their titrations with great care and accuracy.
The volumes of sulphuric acid solution that they each use are tabulated below. Only student A finds the correct concentration of the sulphuric acid. Student B is within 20% but student C is so far out that they know something is wrong. Student C asks for help and is reminded that some solids can contain water of crystallization. Student A uses anhydrous sodium carbonate, but what is x in the formula $Na_2CO_3 \cdot xH_2O(s)$ for students B and C?

	Student A	Student B	Student B
Volume	23.75 cm^3	20.20 cm^3	8.80 cm^3

a) Calculate the exact concentration of the sulphuric acid.

b) Find x for the two different cases of students B and C.

B8 Parts per million

14/17

B8.1 Calculate the ppm by volume of:

a) 20 cm^3 of CO per 40 m^3 of air

b) 0.10 ml of alcohol per 100 ml of blood

c) 5.0 cm^3 of O_3 per 20 m^3 of air

d) 0.0040 cm^3 of C_2H_4 per 1 dm^3 of air

B8.2 Calculate the ppm by mass of:

a) 10 mg of Hg per tonne of water

b) 0.020 g of Mg per kg of $CaCO_3$

c) 50 mg of iron per kg of blood

d) 4.0×10^{-4} moles of arsenic per 1 kg of iron ore

B8.3 Calculate the ppm by number of particles of:

a) 23 mg of sodium in 2 kg of mercury

b) 60 μmol of albumen in 36 cm^3 of water

c) 12 μg of magnesium hydrogen phosphate in 90 μl of water

d) 84 μg of carbon monoxide in 12 dm^3 of air

B8.4 Convert the following concentrations from parts per million (ppm) by mass to mol kg^{-1}.

a) 2500 ppm $CaCO_3$

b) 32.0 ppm NH_3

c) 120 ppm H_2O_2

d) 0.25 ppm Hg

e) 6.0 ppm $CH_3CH_2CH_2COOH$

Chapter C

Gas Laws

$^{24}/_{30}$

C1 Ideal gases

Use the ideal gas equation of state to answer the questions in this section. In SI units, the equation is $pV = nRT$, where $R = 8.31\ \mathrm{J\,K^{-1}\,mol^{-1}}$.

C1.1 a) Give the SI unit of pressure.

b) Give the SI unit of volume.

c) Express 140 °C in kelvin.

d) Give the gradient of a graph of p against T in terms of n, R and V.

C1.2 a) 50 $\mathrm{cm^3}$ of gas at a pressure of 2.5 atm is allowed to expand slowly at constant temperature until it fills a volume of 85 $\mathrm{cm^3}$. Calculate the new pressure of the gas.

b) 20 $\mathrm{dm^3}$ of gas at a pressure of 750 torr is compressed slowly at constant temperature until the pressure reaches 3.0×10^5 torr. Calculate the volume now occupied by the gas.

c) A sealed, rigid container of air at 1 atm pressure falls in temperature from 296 K to 270 K. Find the new pressure inside the container.

d) If the temperature of a gas measured in kelvin is doubled when it was initially at 17 °C, give its new temperature in °C.

e) A canister of gas will explode once the pressure exceeds 40 atm. If the pressure inside is 8.0 atm at 20 °C, find the temperature at which the canister will explode.

f) A sac of gas freely changes its volume to keep its internal pressure equal to atmospheric pressure. If the the sac has a volume of 1.2 $\mathrm{m^3}$ at −10 °C and then warms up to 17 °C with no change in pressure, find its new volume.

C1.3 Complete the table below, using appropriate numbers of significant figures:

p	V	n	T
1.0×10^6 N/m^2	3.2×10^{-4} m^3	0.16 mol	(a)
(b)	120 m^3	80 kmol	286 K
7.5×10^3 Pa	(c)	0.25 mol	77 K
150 kPa	6.0 dm^3	(d)	320 K
(e)	1.8×10^{-3} m^3	0.50 mol	565 °C
245 N/mm^2	252 mm^3	18.0 mmol	(f)

C1.4 A gas cylinder is being filled with argon gas. The gas cylinder has a volume of 24 dm^3 and holds 1 mole of gas at room temperature and pressure.

a) Calculate the amount of gas (in moles) which must be added to raise the pressure in the cylinder from 1 atm to 250 atm. Assume that the volume is constant.

b) If the gas cylinder in part (a) contains a pressure of 250 atm at 20 °C, and is caught in a fire, so that its temperature is raised to 350 °C, calculate the new pressure inside the cylinder.

C1.5 Complete the table below, using the fact that for a fixed amount of gas, pV/T is constant.

p_1	V_1	T_1	p_2	V_2	T_2
1 atm	4 m^3	300 K	(a)	2 m^3	450 K
200 psi	0.6 ft^3	280 K	250 psi	1.2 ft^3	(b)
50 kPa	10 dm^3	25 °C	75 kPa	(c)	−24 °C
(d)	5 m^3	450 °C	1000 atm	5.2 m^3	1350 °C

C1.6 A barometer shows a reading of 29.2 inches of mercury. Use $g = 9.81\ \mathrm{m\,s^{-2}}$ for this question.

a) Given that 1 inch is approximately 2.54 cm, and the density of mercury is 13 590 $\mathrm{kg\,m^{-3}}$, calculate the atmospheric pressure in pascals.

A gasman uses a manometer to check the pressure of gas delivered to a home. The manometer contains water with a density of 1.0 $\mathrm{g\,cm^{-3}}$. The diagram below shows the manometer reading. The scale is marked in decimetres.

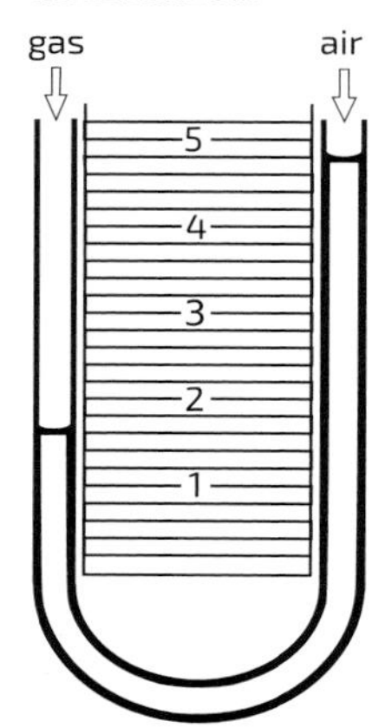

b) Give the reading from the manometer.

c) Calculate the difference between the gas pressure and atmospheric pressure.

d) Calculate the absolute (total) pressure of the gas.

e) Calculate the number of molecules supplied per cubic metre of gas at a temperature of 291 K and under the pressure calculated in part (d).

C1.7 When 2.0 moles of a gas mixture in chemical equilibrium at 1 atm and 296 K is compressed to half its original volume, the temperature rises to 312 K and the pressure rises to 1.7 atm. Calculate the amount of gas mixture present in the smaller volume.

C1.8 A vacuum line is lowered to a pressure of 1.3 kPa at 77 K. Give the number of molecules of gas per $\mathrm{mm^3}$.

C1.9 The endothermic reaction between sodium hydrogencarbonate and ethanoic acid is used to inflate a plastic bag.

a) If the gas produced is at a temperature of 13 °C, and 4.0 g of sodium hydrogencarbonate reacts with excess acid, find the volume of gas produced at a pressure of 101 kPa.

b) If the gas then warms up to a room temperature of 32 °C with no change in pressure, find the new volume of gas.

C2 Density

10/12

C2.1 a) One mole of helium gas has a mass of 4.0 g and a volume of 24 dm^3. Calculate its density in $kg\,m^{-3}$.

b) Calculate the density of an ideal gas of molar mass 30.0 $g\,mol^{-1}$ at 101 kPa and 295 K.

c) Assuming a molar volume of 24 $dm^3\,mol^{-1}$, calculate the density of dry air (78% N_2, 21% O_2, 1% Ar) to 2 s.f.

C2.2 Does the presence of water vapour in air make the density higher or lower than dry air?

C2.3 An unknown gas is found to have a density of 0.702 $kg\,m^{-3}$ at 291 K and 99.8 kPa. Find its molar mass.

C2.4 A gas, $W(CO)_n$, has a density of 13.2 $kg\,m^{-3}$ at 320 K and 9.95×10^4 Pa. Find n.

C2.5 Find the temperature at which the density of carbon dioxide falls to 1.10 $g\,dm^{-3}$ at atmospheric pressure, 101.3 kPa.

C2.6 Find the pressure at 280 K at which oxygen has a density of 40.0 $kg\,m^{-3}$.

C2.7 Find the volume occupied by 1.00 g of methane at 293 K and 105 kPa.

C2.8 Give the concentration in $mol\,dm^{-3}$ of a solution of an ideal gas in water that has the same average distance between its molecules as in the pure gas at 273 K and at 100 Pa.

C2.9 The density of chlorine at its critical point is 573 $kg\,m^{-3}$. Its critical temperature is 417 K, and its critical pressure is 7.71 MPa. How many times denser is chlorine at its critical point than it would be if it were an ideal gas under the same conditions?

C2.10 A cylindrical container of internal height 1.25 m and internal radius 0.10 m is filled with hydrogen to a pressure of 2.0×10^7 $N\,m^{-2}$ at 290 K. Find the mass of hydrogen stored in the cylinder.

4/5

C3 Non-ideal gases

C3.1 Under what conditions is the ideal gas law most accurate?

________________ pressure

________________ temperature

C3.2 Carbon monoxide is denser than both nitrogen and ethene. This implies that carbon monoxide has ________________ intermolecular forces that make the molar volume ________________ than predicted by the ideal gas equation, so that there are ________________ molecules per cm^3. At a given density and temperature, carbon monoxide exerts a ________________ pressure than ethene or nitrogen.

C3.3 The van der Waals equation of state is:

$$\left(p + \frac{n^2 a}{V^2}\right)(V - nb) = nRT,$$

where a and b are constants.

a) Find the temperature of 3.7 mol of gas occupying 0.0042 m^3 at 170 kPa if $a = 0.052\ \mathrm{N\,m^4\,mol^{-2}}$ and $b = 1.1 \times 10^{-4}\ \mathrm{m^3\,mol^{-1}}$.

b) Find the pressure exerted by 2.0 mol of gas at 400 K in a volume of 7.3 dm^3 if $a = 0.37 \times 10^4\ \mathrm{N\,dm^4\,mol^{-2}}$ and $b = 4.2 \times 10^{-2}\ \mathrm{dm^3\,mol^{-1}}$.

c) Find b for 1 mol of a gas at a pressure of 120 kPa occupying $2.0 \times 10^{-2}\ \mathrm{m^3}$ at 290 K if $a = 1.2\ \mathrm{N\,m^4\,mol^{-2}}$.

Chapter D

Atomic Structure

D1 Atomic structure

24/30

Give the missing numbers or letters in each of the following ground state electron configurations. A missing number or letter is indicated by a "**?**".

D1.1
- a) Be $1s^2\,2s^{?}$
- b) N $1s^2\,2s^2\,?p^3$
- c) Ne $1s^2\,2s^2\,2?^6$

D1.2
- a) Mg $[Ne]\,?s^2$
- b) Si $[Ne]\,3s^2\,3?^2$
- c) S $[Ne]\,3s^2\,3p^{?}$

D1.3
- a) K $[Ar]\,?s^1$
- b) Sc $[Ar]\,3d^{?}\,4s^{?}$
- c) Cr $[Ar]\,3d^{?}\,4s^{?}$
- d) Co $[Ar]\,3d^{?}\,4s^{?}$
- e) Cu $1s^2\,2s^2\,2p^6\,3s^2\,3p^6\,3d^{?}\,4s^{?}$

D1.4
- a) H^- $1s^{?}$
- b) O^{2-} $1s^2\,2s^2\,2p^{?}$
- c) Na^+ $1s^2\,2s^2\,?p^6$
- d) Al^{3+} $1s^2\,2s^2\,?p^{?}$

D1.5
- a) Cl^- $[Ne]\,3s^2\,3?^6$
- b) Ca^{2+} $[Ne]\,3s^2\,3p^{?}$

D1.6
- a) Ti^{3+} $[Ar]\,??^1$
- b) Fe^{2+} $[Ar]\,3d^{?}$
- c) Ni^{2+} $[Ar]\,3d^{?}$
- d) Cu^+ $[Ar]\,3d^{?}$
- e) Zn^{2+} $1s^2\,2s^2\,2p^6\,3s^2\,3p^6\,3d^{?}$

D1.7 Give the chemical symbols for the atoms with the following ground state electron configurations:

- a) $[Ne]\,3s^1$
- b) $[Ar]\,3d^5\,4s^2$
- c) $1s^2\,2s^2\,2p^6\,3s^2\,3p^6\,3d^8\,4s^2$
- d) $[Ar]\,3d^{10}\,4s^2$

e) $1s^2\,2s^2\,2p^6\,3s^2\,3p^6\,3d^{10}\,4s^2\,4p^6\,4d^{10}\,4f^{14}\,5s^2\,5p^6\,5d^{10}\,6s^2\,6p^5$

D1.8 An ion of nickel is found to have the ground state electron configuration $1s^2\,2s^2\,2p^6\,3s^2\,3p^6\,3d^7$ in the gas phase. Give the charge on the ion.

D1.9 Give the number of electrons in the third shell of an atom of vanadium in the ground state.

D1.10 A 1^+ ion in an excited state is found to have an electron configuration $1s^2\,2s^1\,2p^6\,3s^2\,3p^6\,3d^6\,4s^2\,4p^1$ in the gas phase. Name the element whose ion this is.

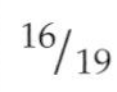

D2 Atomic orbitals

D2.1 a) Give the number of f-orbitals that comprise the 5f subshell.

b) Give the maximum number of electrons that can occupy a single orbital.

c) Give the maximum number of electrons that can occupy the second shell.

D2.2 a) Give the maximum number of unpaired electrons that can occupy the 3d subshell.

b) Give the number of unpaired electrons in the ground state of an oxygen atom.

c) Give the number of paired electrons in the ground state of the Na^+ ion.

D2.3 Identify the subshell to which the following orbitals belong:

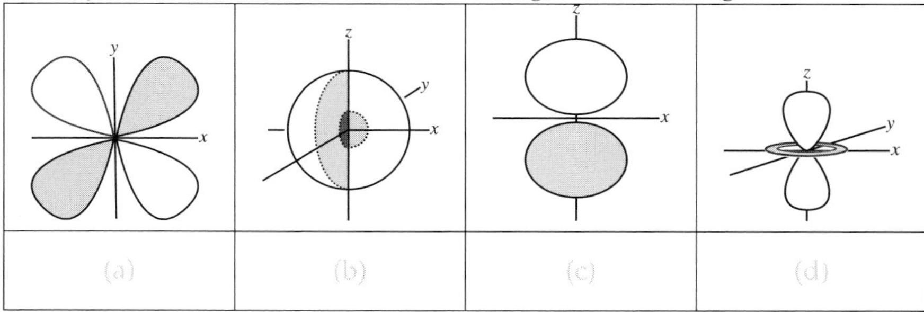

The free Isaac3D virtual reality app for Android smart phones shows atomic orbitals in 3-D on any simple VR device, for instance Googlecardboard (vr.google.com/cardboard/).

D2.4 Give the number of occupied orbitals in the ground state of a vanadium atom.

D2.5 Give the highest occupied subshell in the Ti^{4+} ion.

D2.6 Give the number of electrons in p-orbitals in the ground state of the Bi^{3+} ion.

D2.7 a) Give the principal quantum number associated with the second shell.

b) Give the orbital angular momentum quantum number, l, associated with an s-subshell.

c) An electron in the third shell has a magnetic quantum number of 2. Give the subshell in which it is found.

D2.8 Give the sum of the m_s quantum numbers for all of the electrons in an atom of zinc in the ground state.

D2.9 Give the sum of the m_l quantum numbers for all of the electrons in an atom of zinc in the ground state.

D2.10 An electron in an orbital has quantum numbers, $n = 4$, $l = 1$, $m_l = 0$, $m_s = -\frac{1}{2}$. Give the subshell in which this electron is found.

D3 Atomic and ionic radii and ionization energy

$^{17}/_{21}$

D3.1 There are trends evident in atomic and ionic radii. Ionization energies also show trends. Complete the following sentences with the words 'increase' or 'decrease', to indicate what happens to the radii of the atoms or ions [(a)–(f)], or to the ionization energies [(g)–(i)].

a) Going along a period from left to right, the atomic radii...

b) Going down a group, the atomic radii...

c) As successive electrons are removed from the same atom/ion, the radii...

d) The radii of ions of the same charge, on descending a group...

e) As successive electrons are added to one atom to make increasingly negative ions, the radii...

f) Going along a period from left to right, the radii of isoelectronic species...

g) Along a period from left to right, the first ionization energies generally...

h) Going down a group, the first ionization energies...

i) Successive ionization energies for the same element...

D3.2 a) Which would have the smallest radius in the set [Na Mg Al]?

b) Which would have the smallest radius in the set [Na^+ Mg^{2+} Al^{3+}]?

c) Which would have the smallest radius in the set [B Al Ga In Tl]?

d) Which would have the largest radius in the set [Si^{4-} P^{3-} S^{2-} Cl^-]?

e) Which would have the smallest radius in the set [Ti^{4+} Zr^{4+} Hf^{4+} Rf^{4+}]?

f) Which would have the largest radius in the set [Fe Fe^{2+} Fe^{3+} Fe^{2-}]?

D3.3 An element has its first to fifth ionization energies in $kJ\,mol^{-1}$ listed as: 578, 1817, 2745, 11578, 14831. Give the group number in the periodic table that corresponds to this element.

D3.4 The first three ionization energies of chlorine, in $kJ\,mol^{-1}$, are: 1251, 2297, 3822. Give the enthalpy change in $kJ\,mol^{-1}$ for the gas-phase process: $Cl \longrightarrow Cl^{2+} + 2\,e^-$.

D3.5 The first four ionization energies of titanium, in $kJ\,mol^{-1}$, are: 658, 1310, 2653, 4175. Give the enthalpy change in $kJ\,mol^{-1}$ for the gas-phase process: $2\,Ti \longrightarrow 2\,Ti^{2+} + 4\,e^-$.

D3.6 The two ionization energies of helium are 2372 $kJ\,mol^{-1}$ and 5251 $kJ\,mol^{-1}$. Give the enthalpy change occuring, in $kJ\,mol^{-1}$, when a free alpha particle combines with two free electrons to produce an atom of helium.

D3.7 Given that the first ionization energies of arsenic and bromine are 947 and 1140 $kJ\,mol^{-1}$ respectively, estimate the first ionization energy of selenium in $kJ\,mol^{-1}$.

D3.8 The first, second and fourth ionization energies of krypton are 1351, 2368 and 5070 $kJ\,mol^{-1}$ respectively. Estimate the third ionization energy of krypton in $kJ\,mol^{-1}$.

D3.9 Ultraviolet light of wavelength 215 nm ejects electrons with a maximum kinetic energy of 0.15 eV from lanthanum atoms in the gas phase. Calculate the first ionization energy of lanthanum in $kJ\,mol^{-1}$.

D4 Isotopes

D4.1 Name the isotopes with the following numbers protons and neutrons in their nuclei, e.g. 2 protons and 2 neutrons gives the answer helium−4.

a) 1 proton and 2 neutrons

b) 5 protons and 6 neutrons

c) 15 protons and 16 neutrons

d) 18 protons and 22 neutrons

e) 27 protons and 33 neutrons

f) 35 protons and 44 neutrons

g) 38 protons and 52 neutrons

h) 55 protons and 82 neutrons

i) 90 protons and 142 neutrons

j) 95 protons and 146 neutrons

D4.2 Complete the table to show the numbers of protons and neutrons in each isotope:

	ISOTOPE	# PROTONS	# NEUTRONS
(a)	Carbon−12		6
(b)	Carbon−13		
(c)	Technetium−99	43	
(d)	Iodine−131		
(e)	Polonium−210		
(f)	Uranium−233		
(g)	Rutherfordium−260		

D4.3 Complete the following table by filling any blank cell and any missing symbol indicated by a "**?**":

	Symbol	# PROTONS	# NEUTRONS	# ELECTRONS
(a)	$^{23}_{11}Na$		12	
(b)	$^{40}_{19}K$			
(c)	$^{25}_{12}Mg^{2+}$	12		
(d)	$^{81}_{35}Br^{-}$			
(e)	$^{58}_{26}Fe^{3+}$			
(f)	$^{18}_{8}O^{2-}$			
(g)	$^{206}_{82}\mathbf{?}$			82
(h)	$^{239}_{93}\mathbf{?}$			93

D4.4 Complete the missing numbers and symbols (each indicated by a "**?**") in the following decay equations:

a) $^{40}_{19}K \longrightarrow {}^{40}_{20}\mathbf{?} + {}^{0}_{-1}\beta + {}^{0}_{0}\overline{\nu}$

b) $^{210}_{84}Po \longrightarrow {}^{\mathbf{?}}_{82}Pb + {}^{4}_{2}\alpha$

c) $^{\mathbf{?}}_{53}I \longrightarrow {}^{131}_{54}Xe + {}^{\mathbf{?}}_{-1}\beta + {}^{0}_{0}\overline{\nu}$

d) $^{65}_{\mathbf{?}}Zn + {}^{0}_{\mathbf{?}}e \longrightarrow {}^{65}_{\mathbf{?}}Cu + {}^{0}_{0}\nu$

e) $^{\mathbf{?}}_{19}K \longrightarrow {}^{40}_{18}Ar + {}^{0}_{\mathbf{?}}\beta^{+} + {}^{0}_{0}\nu$

f) $^{\mathbf{?}}_{92}U \longrightarrow {}^{143}_{56}\mathbf{?} + {}^{90}_{\mathbf{?}}Kr + 3\,{}^{1}_{0}n$

Chapter E

Electronic Spectroscopy

$^{24}/_{29}$

The ionization energy of hydrogen is approximately 1314 kJ mol^{-1}.

E1.1 a) Express the energy required to ionize a single hydrogen atom in

i. J

ii. eV

b) Calculate the 2nd ionization energy of helium, in kJ mol^{-1}.

c) The energy required to promote an electron from the 1st shell to the 2nd in an atom of hydrogen is 1.64 aJ. Calculate the energy required to ionize an excited hydrogen atom with electron configuration $2p^1$.

d) Calculate the energy required to promote an electron from the 2nd shell to the 3rd shell in a hydrogen atom.

E1.2 Give the frequency of a photon that

a) has an energy of 4.0×10^{-17} J;

b) could just ionize an atom of hydrogen in its ground state;

c) is emitted when an electron in a hydrogen atom falls from the 2nd shell to the 1st;

d) is absorbed when an electron in a hydrogen atom is promoted from the 1st shell to the 3rd;

e) is absorbed when an electron in an He^+ ion is promoted from the 1st shell to the 2nd.

E1.3 Give the wavelength of a photon in a vacuum that

a) has an energy of 2.6×10^{-15} J;

b) could just ionize an atom of hydrogen in its ground state;

c) is absorbed when an electron in a hydrogen atom is promoted from the 1st shell to the 2nd;

d) is emitted when an electron in a hydrogen atom falls from the 4th shell to the 1st;

e) is absorbed when an electron in an Li^{2+} ion is promoted from the 2nd shell to the 3rd.

E1.4 In the absorption spectrum of hydrogen, absorbed frequencies of light are detected as [dark/coloured] lines in the spectrum. The lines appear in [distinct/overlapping] series. The lines in each series converge at [high/low] frequency. Lines in the same series are caused by transitions [from/to] the same [upper/lower] shell. The frequencies absorbed from the visible region in the atomic absorption spectrum of hydrogen are due to transitions [from/to] the [1st/2nd/3rd] shell.

E1.5 Indicate the transitions that could be responsible for lines in the emission spectrum of a hydrogen atom.

A) $2p \longrightarrow 1s$

B) $2s \longrightarrow 1s$

C) $3p \longrightarrow 1s$

D) $3d \longrightarrow 2p$

E) $1s \longrightarrow 4p$

F) $2s \longrightarrow 3p$

G) $4d \longrightarrow 2p$

H) $4f \longrightarrow 2p$

I) $3s \longrightarrow 3p$

J) $4f \longrightarrow 4d$

K) $4f \longrightarrow 3d$

L) $35s \longrightarrow 2p$

M) $35p \longrightarrow 2s$

N) $12s \longrightarrow 1s$

Chapter F

Enthalpy Changes

F1 Calorimetry

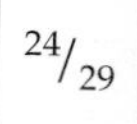

Specific heat capacity of water = 4.18 $J\,g^{-1}\,K^{-1}$.

F1.1

a) Calculate the heat capacity of an object with mass 1.80 kg and specific heat capacity 0.32 $J\,g^{-1}\,K^{-1}$.

b) Calculate the heat capacity of a calorimeter if its temperature is raised 2.5 K by 35 kJ of heat.

c) Calculate the expected increase in temperature when 2.4 kJ of heat is transferred to a calorimeter of heat capacity 720 $J\,K^{-1}$.

d) Calculate the heat required to raise the temperature of a calorimeter of heat capacity 1.6 $kJ\,K^{-1}$ by 3.8 °C.

e) Calculate the specific heat capacity of a calorimeter if it has a mass of 375 g and its temperature is raised 4.2 K by 2160 J of heat.

f) Calculate the heat required to raise the temperature of 3.14 kg of water by 12.2 K.

g) Calculate the mass of water whose temperature is raised through 16.0 K by 6.7 kJ of heat.

h) A calorimeter consists of 140 g of aluminium and 300 g of water. 6100 J of heat raises its temperature by 4.42 K. Calculate:

   i. its heat capacity;
   ii. the specific heat capacity of aluminium.

F1.2 Complete each row in the table below:

Initial T / °C	Final T / °C	Mass / kg	Specific heat capacity / J kg^{-1} K^{-1}	Heat gained / kJ
20.0	24.0	(a)	4180	20.9
18.0	19.2	2.50	(b)	12.0
16.3	(c)	5.00	450	348
17.90	18.04	0.40	840	(d)
19.50	16.00	1.60	4180	(e)
(f)	32.6	2.20	4180	62.5
−2.0	14.5	(g)	280	3.70

F1.3 The enthalpy change of combustion of naphthalene is −5156 kJ mol^{-1}. Its molar mass is 128.2 g mol^{-1}. Calculate the temperature change expected when 1.00 mmol is burnt in excess oxygen in a calorimeter containing 4.0 kg of water.

F1.4 The enthalpy change of combustion of decane, $C_{10}H_{22}$, is −6778 kJ mol^{-1}. Calculate the mass required to raise the temperature of 450 g of water by 80 °C when burnt completely, with no heat losses from the water.

F1.5 Complete combustion of 0.020 mol of ethane, with a standard enthalpy of combustion of −1410.8 kJ mol^{-1} raises the temperature of the water in an insulated calorimeter from 17.4 °C to 22.4 °C. Calculate the mass of the water in the calorimeter.

F1.6 Calculate the enthalpy of combustion of propyne, C_3H_4, given that complete combustion of 65 mg of propyne raises the temperature of 800 g of water from 20.15 °C to 21.09 °C.

F1.7 25.0 cm^3 of sulphuric acid at 1.00 mol dm^{-3} and 19.10 °C is placed in an insulated polystyrene cup. When 25.0 cm^3 of sodium hydroxide at 2.00 mol dm^{-3} and 19.10 °C is added, the temperature rises to 32.45 °C.

Assuming that no heat is lost, that the specific heat capacity of water may be used, and that the solutions have a density of 1.00 $g\,cm^{-3}$ at 19.10 °C, find the enthalpy change of the reaction per mole of water produced by neutralisation.

F1.8 30.0 cm^3 of ethanoic acid at 1.60 $mol\,dm^{-3}$ and 18.65 °C is placed in an insulated polystyrene cup. When 40.0 cm^3 of sodium hydroxide at 1.00 $mol\,dm^{-3}$ and 18.65 °C is added, the temperature rises to 25.80 °C. Assuming that no heat is lost, that the specific heat capacity of water may be used, and that the solutions have a density of 1.00 $g\,cm^{-3}$ at 18.65 °C, find the enthalpy change of the reaction per mole of water produced by neutralisation.

F1.9 When 5.0 g of ammonium nitrate dissolves in 100 g of water, the temperature of the water drops from 18 °C to 14 °C. Calculate the enthalpy of solution of ammonium nitrate in $kJ\,mol^{-1}$.

a) Write down the formula of ammonium nitrate.

b) Calculate the formula mass of ammonium nitrate.

c) Calculate the number of moles of ammonium nitrate in 5.0 g.

d) Calculate the heat lost from the 100 g of water.

e) Calculate the heat lost per mole of ammonium nitrate.

f) Give the enthalpy of solution of ammonium nitrate.

F1.10 The enthalpies of combustion of three fuels are shown below:

Fuel	$\Delta_c H$ / $kJ\,mol^{-1}$
CH_4	−890.3
C_3H_8	−2219.2
C_4H_{10}	−2876.5

a) Which gives out most heat per gram?

b) Which gives out most heat per mole?

c) Which gives out most heat per cubic foot?

16/19

F2 Bond enthalpies

F2.1 Use mean bond enthalpies in the table below to calculate the reaction enthalpies for the following reactions in the gas phase:

a) $H_2 + Cl_2 \longrightarrow 2\,HCl$

b) $H_2 + I_2 \longrightarrow 2\,HI$

c) $CH_4 + Cl_2 \longrightarrow CH_3Cl + HCl$

d) $C_2H_4 + H_2 \longrightarrow C_2H_6$

e) $C_2H_2 + 2\,H_2 \longrightarrow C_2H_6$

f) $C_4H_{10} \longrightarrow C_2H_6 + C_2H_4$

g) $C_2H_4 + HCl \longrightarrow CH_3CH_2Cl$

h) $\frac{1}{2}Cl_2 \longrightarrow Cl$

i) $C_2H_4 + I_2 \longrightarrow C_2H_4I_2$

j) $CH_2CHCHCH_2 + 2\,H_2 \longrightarrow C_4H_{10}$

MEAN BOND ENTHALPIES (in $kJ\,mol^{-1}$)

H–H	436	C–Cl	327
C–H (average)	413	I–I	151.2
C–H (in CH_4 and CH_3X)	435	H–I	298.3
C–C	347	C–I	228
C=C	612	Br–Br	193
C≡C	838	H–Br	366.3
H–Cl	432	N≡N	945.4
Cl–Cl	243.4	O=O	498.3

F2.2 Use the reaction enthalpies given (for the gas phase reaction), and the bond enthalpies above, to find the bond enthalpy requested:

a) $I_2 + Cl_2 \longrightarrow 2\,ICl$
$\Delta_r H^\circ = -70.2\ kJ\,mol^{-1}$, find E(I–Cl)

b) $CH_4 + Br_2 \longrightarrow CH_3Br + HBr$
$\Delta_r H^\circ = -28.3\ kJ\,mol^{-1}$, find E(C–Br)

c) $N_2 + 3\,H_2 \longrightarrow 2\,NH_3$
$\Delta_r H^\circ = -92.2\ kJ\,mol^{-1}$, find E(N–H)

d) $O_2 + 2\,H_2 \longrightarrow 2\,H_2O$
$\Delta_r H^\circ = -483.6$ kJ mol^{-1}, find E(O–H)

e) $3\,O_2 \longrightarrow 2\,O_3$
$\Delta_r H^\circ = 285.4$ kJ mol^{-1}, find E(O–O) in ozone

F2.3 Use some of the bond enthalpies below (in kcal mol^{-1}) to calculate the enthalpy changes for the reactions (in the gas phase):

a) $C_2H_4 + O_2 \longrightarrow 2\,CH_2O$

b) $CO + H_2O \longrightarrow CO_2 + H_2$

c) $CH_2O \longrightarrow CO + H_2$

MEAN BOND ENTHALPIES (in kcal mol^{-1})

C=C	146	C≡O	258
O=O	119	O–H	111
C–H	99	H–H	104
C=O	178		

F2.4 Given that the bond dissociation energy of H–H is 4.53 eV, D–D is 4.59 eV, and the energy change on reaction $H_2 + D_2 \longrightarrow 2\,HD$ is +0.02 eV, find the bond energy of H–D.

F3 Formation and combustion enthalpies

22/27

Data (all in kJ mol^{-1}):

	$\Delta_f H^\circ$		$\Delta_c H^\circ$
CH_4(g)	−74.8	C_6H_6(l)	−3267.4
CCl_4(l)	−129.6	H_2(g)	−285.8
HCl(g)	−92.3	C_6H_{12}(l)	−3919.5
$TiCl_4$(l)	−804.2	C_2H_2(g)	−1300.8
$TiCl_3$(s)	−720.9	C_2H_6(g)	−1559.7
PCl_3(l)	−319.7	C_2H_5OH(l)	−1367.3
PCl_5(s)	−443.5	C_2H_4(g)	−1410.8
$POCl_3$(l)	−597.1	CH_3COOH(l)	−874.1
GeO(s)	−212.1	C_6H_{14}(l)	−4163.0
GeO_2(s)	−551.0	$CH_3COOC_2H_5$(l)	−2237.9
NH_3(g)	−46.1	CO(g)	−283.0
TiO_2(s)	−939.7	Mg(s)	−601.7

F3.1 Use standard enthalpies of formation to calculate the reaction enthalpies for the following reactions:

a) $2\,TiCl_3(s) + Cl_2(g) \longrightarrow 2\,TiCl_4(l)$

b) $PCl_3(l) + Cl_2(g) \longrightarrow PCl_5(s)$

c) $2\,PCl_3(l) + O_2(g) \longrightarrow 2\,POCl_3(l)$

d) $CH_4(g) + 4\,Cl_2(g) \longrightarrow CCl_4(l) + 4\,HCl(g)$

e) $2\,GeO(s) \longrightarrow Ge(s) + GeO_2(s)$

f) $GeO(s) + PCl_3(l) \longrightarrow Ge(s) + POCl_3(l)$

g) $PCl_5(s) + 2\,TiCl_3(s) \longrightarrow PCl_3(l) + 2\,TiCl_4(l)$

h) $20\,Ti(s) + 12\,PCl_5(s) \longrightarrow 20\,TiCl_3(s) + 3\,P_4(s)$

F3.2 Use standard enthalpies of combustion to calculate the reaction enthalpies for the following reactions:

a) $C_2H_2(g) + 2\,H_2(g) \longrightarrow C_2H_6(g)$

b) $C_6H_6(l) + 3\,H_2(g) \longrightarrow C_6H_{12}(l)$

c) $3\,C_2H_2(g) \longrightarrow C_6H_6(l)$

d) $C_2H_4(g) + H_2O(l) \longrightarrow C_2H_5OH(l)$

e) $C_2H_5OH(l) + O_2(g) \longrightarrow CH_3COOH(l) + H_2O(l)$

f) $C_6H_{14}(l) \longrightarrow C_2H_6(g) + 2\,C_2H_4(g)$

g) $C_2H_5OH(l) + CH_3COOH(l) \longrightarrow CH_3COOC_2H_5(l) + H_2O(l)$

h) $2\,C_2H_2(g) + 2\,H_2O(l) + O_2(g) \longrightarrow 2\,CH_3COOH(l)$

F3.3 Use enthalpies of formation and combustion to calculate the reaction enthalpy for the reaction:
$Ge(s) + 2\,H_2O(l) \longrightarrow GeO_2(s) + 2\,H_2(g)$

F3.4 Use the reaction enthalpies given, and the combustion or formation enthalpies above to find the requested enthalpy change in each case:

a) $NH_3(g) + HCl(g) \longrightarrow NH_4Cl(s)$
$\Delta_r H^\circ = -176\ \text{kJ mol}^{-1}$, find $\Delta_f H^\circ$ of $NH_4Cl(s)$

b) $TiCl_4(l) + 2\,Mg(s) \longrightarrow 2\,MgCl_2(s) + Ti(s)$
$\Delta_r H^\circ = -478.4\ \text{kJ mol}^{-1}$, find $\Delta_f H^\circ$ of $MgCl_2(s)$

c) $CH_3COOCOCH_3(l) + H_2O(l) \longrightarrow 2\,CH_3COOH(l)$
$\Delta_r H^\circ = -46\ \text{kJ mol}^{-1}$, find $\Delta_c H^\circ$ of $CH_3COOCOOH_3(l)$

d) $4\,C_2H_2(g) \longrightarrow C_6H_5CHCH_2(l)$
$\Delta_r H^\circ = -808.2$ kJ mol^{-1}, find $\Delta_c H^\circ$ of $C_6H_5CHCH_2$

e) $4\,Al(s) + 3\,GeO_2(s) \longrightarrow 2\,Al_2O_3(s) + 3\,Ge(s)$
$\Delta_r H^\circ = -1698.4$ kJ mol^{-1}, find $\Delta_f H^\circ$ of $Al_2O_3(s)$

f) $Fe_2O_3(s) + 3\,CO(g) \longrightarrow 2\,Fe(s) + 3\,CO_2(g)$
$\Delta_r H^\circ = -24.8$ kJ mol^{-1}, find $\Delta_f H^\circ$ of Fe_2O_3

g) $3\,CuO(s) + 2\,NH_3(g) \longrightarrow 3\,Cu(s) + N_2(g) + 3\,H_2O(l)$
$\Delta_r H^\circ = -293.3$ kJ mol^{-1}, find $\Delta_f H^\circ$ of $CuO(s)$

h) $2\,PCl_5(s) + 8\,H_2O(l) \longrightarrow 2\,H_3PO_4(s) + 10\,HCl(g)$
$\Delta_r H^\circ = -307.6$ kJ mol^{-1}, find $\Delta_f H^\circ$ of $H_3PO_4(s)$

i) $Ga_2O_3(s) + 3\,Mg(s) \longrightarrow 2\,Ga(s) + 3\,MgO(s)$
$\Delta_r H^\circ = -716.1$ kJ mol^{-1}, find $\Delta_c H^\circ$ of Ga

j) $TiCl_4(l) + 2\,H_2O(l) \longrightarrow TiO_2(s) + 4\,HCl(aq)$
$\Delta_r H^\circ = -232.3$ kJ mol^{-1}, find $\Delta_{sol} H^\circ$ of HCl(g)

F4 Born-Haber cycles

15/18

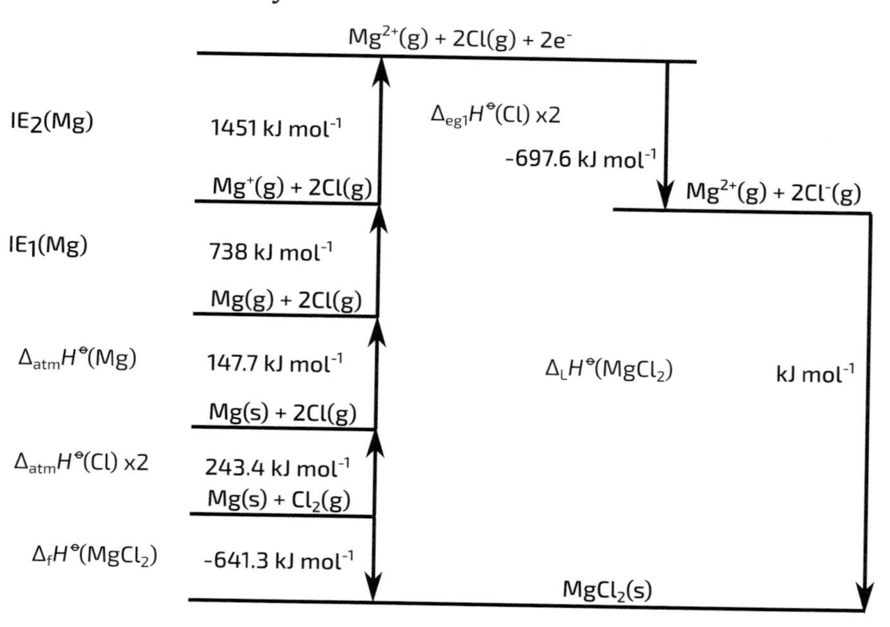

F4.1 Calculate the lattice enthalpy of magnesium chloride.

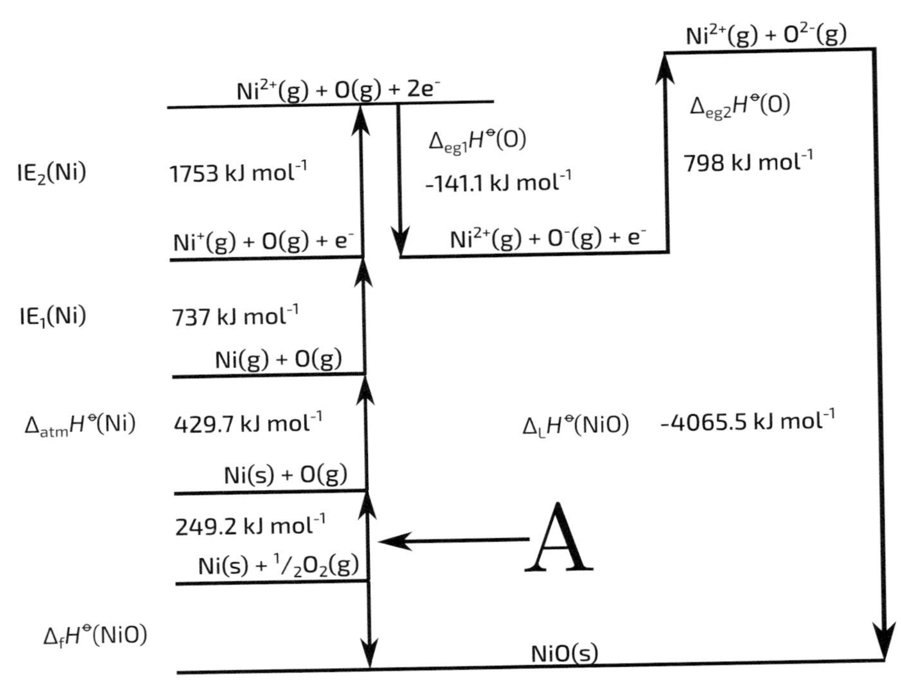

F4.2 a) Use the Born-Haber cycle above to find the enthalpy of formation of nickel(II) oxide.

b) The enthalpy change labelled A is the ________________ enthalpy of oxygen.

F4.3 Fill in the missing value in each column of the table below, assuming that a Born-Haber cycle may be used to calculate the lattice enthalpy in the usual way.

COMPOUND	LiBr	Na_2O	CaF_2	Cr_2O_3	Hg_2Cl_2
$\Delta_f H^{\circ}$	−351.2	−414.2	−1219.6	−1 139.7	−265.2
$\Delta_{atm} H^{\circ}$ (metal)	159.4	107.3	178.2	(a)	61.3
$\Delta_{atm} H^{\circ}$ (non-metal)	111.9	249.2	(b)	249.2	121.7
$\Delta_{i.e.1} H^{\circ}$	520	(c)	590	653	1007
$\Delta_{i.e.2} H^{\circ}$	7298	4563	1145	1592	1810
$\Delta_{i.e.3} H^{\circ}$	11815	6913	4912	2987	3300
$\Delta_{eg1} H^{\circ}$	−324.6	−141.1	−328	−141.1	(d)
$\Delta_{eg2} H^{\circ}$	N/A	798	N/A	798	N/A
$\Delta_L H^{\circ}$	(e)	−2526.9	−2634.8	−15115.2	−1947.6

All values are in kJ mol^{-1}.

$\Delta_f H^{\circ}$ = standard enthalpy of formation
$\Delta_{atm} H^{\circ}$ (metal) = standard enthalpy of atomisation of the metal
$\Delta_{atm} H^{\circ}$ (non-metal) = standard enthalpy of atomisation of the non-metal
$\Delta_{i.e.1} H^{\circ}$ = first ionization energy of the metal
$\Delta_{i.e.2} H^{\circ}$ = second ionization energy of the metal
$\Delta_{i.e.3} H^{\circ}$ = third ionization energy of the metal
$\Delta_{eg1} H^{\circ}$ = first standard electron gain enthalpy of the non-metal
$\Delta_{eg2} H^{\circ}$ = second standard electron gain enthalpy of oxygen
$\Delta_L H^{\circ}$ = standard lattice enthalpy

F4.4 Within the ionic model, lattice enthalpies in kJ mol^{-1} may be estimated using the equation

$$\Delta_L H^\circ = \frac{C \cdot z_+ \cdot z_- \cdot \nu}{(r_+ + r_-)} - 2.5\nu,$$

where:

- C is a constant approximately equal to 105000 units;
- z_+ and z_- are the signed charges on the cation and anion respectively, in units of e;
- ν is the number of ions in the formula (e.g. 3 for MgI_2);
- r_+ and r_- are the radii of the ions in pm;
- the -2.5ν term corrects for the difference between internal energy and enthalpy.

a) The table below shows the radii for certain ions.

Ion	Li^+	Na^+	Ca^{2+}	Cr^{3+}	Hg^+	O^{2-}	F^-	Cl^-	Br^-
Radius/pm	74	102	100	62	158	140	133	180	195

Estimate the values of $\Delta_L H^\circ$ for the following compounds using the equation given:

i. LiBr
ii. Na_2O
iii. CaF_2
iv. Cr_2O_3
v. Hg_2Cl_2

b) For which compound is the ionic model a poor approximation?

F4.5 Rank the following sets of compounds in order of increasing magnitude of lattice enthalpy, as predicted using the ionic model:

a) KI, LiCl, NaBr
b) MgO, NaF, GaAs
c) $TiCl_2$, $TiCl_3$, $TiCl_4$

F4.6 The molar enthalpy change on decomposition of calcium carbonate into calcium oxide and carbon dioxide is 178 kJ mol^{-1}. Given the lattice enthalpies of $CaCO_3$ (−2799 kJ mol^{-1}) and CaO (−3396 kJ mol^{-1}), find the enthalpy change when bonding an oxide ion to carbon dioxide in the gas phase.

Chapter G

Entropy

G1 Absolute entropy

$^8/_{10}$

Use the following standard molar entropy values in $J\,K^{-1}\,mol^{-1}$ to help answer the questions in this section.

$H_2O(l)$	69.9	$HCl(g)$	186.8	$NaCl(s)$	72.1
$H_2O(g)$	188.7	$Cl_2(g)$	223.1	$ZnCl_2(s)$	111.5
$H_2(g)$	130.7	$H_2SO_4(l)$	156.9	$Zn(s)$	41.6
$Na(s)$	51.2	$Zn(g)$	150.0	$NaHSO_4(s)$	113.0
$O_2(g)$	205.2	$CO_2(g)$	213.6	C(s) graphite	5.7

G1.1 Calculate the entropy of 1.00 kg of liquid water.

G1.2 Calculate the entropy of 3.00 mol of steam.

G1.3 Calculate the entropy of 2.00 dm^3 of hydrogen at room temperature and pressure.

G1.4 Calculate the entropy of 355.0 g of chlorine.

G1.5 Calculate the entropy of 9.81 g of conc. sulfuric acid.

G1.6 Calculate the entropy of 1.00 kg of solid zinc.

G1.7 Calculate the entropy of a sample of graphite containing 2.0×10^{19} atoms.

G1.8 Calculate the mass of sodium chloride that has standard entropy of 100 $J\,K^{-1}$.

G1.9 Calculate the total entropy in 1.00 mol of solid zinc and 1.00 mol of chlorine.

G1.10 Calculate the total entropy in 250 cm^3 of hydrogen and 500 cm^3 of chlorine held separately at room temperature and pressure.

12/15

G2 Entropy changes

G2.1 Calculate the standard entropy change per mole for the following reactions:

a) $H_2O(l) \longrightarrow H_2O(g)$

b) $Zn(s) + Cl_2(g) \longrightarrow ZnCl_2(s)$

c) $H_2(g) + Cl_2(g) \longrightarrow 2\,HCl(g)$

d) $NaCl(s) + H_2SO_4(l) \longrightarrow NaHSO_4(s) + HCl(g)$

e) $Zn(s) + 2\,HCl(g) \longrightarrow ZnCl_2(s) + H_2(g)$

G2.2 Calculate the standard entropy change when...

a) 2.50 mol of solid zinc chloride decomposes into its gaseous elements.

b) 2.0 g of sodium reacts fully with chlorine gas.

c) 10.0 mg of zinc vapour sublimes onto a surface.

d) 40.0 m^3 (at RTP) of steam condenses.

e) 200 cm^3 of water (1.00 $g\,cm^{-3}$) is electrolysed.

G2.3 The decomposition of hydrogen peroxide has a standard entropy change of 62.9 $J\,K^{-1}\,mol^{-1}$. Find the standard molar entropy of hydrogen peroxide.

G2.4 The combustion of methane has a standard molar entropy change of −243.2 $J\,K^{-1}\,mol^{-1}$. Calculate the standard molar entropy of methane.

G2.5 The oxidation of zinc in air has a standard molar entropy change of −100.6 $J\,K^{-1}\,mol^{-1}$. Calculate the standard molar entropy of zinc oxide.

G2.6 The standard molar entropy change on formation of one mole of ethanol from its elements in their reference states is −345.6 $J\,K^{-1}\,mol^{-1}$. Find the standard molar entropy of ethanol.

G2.7 The standard molar entropy change on $C(\text{s graphite}) + \frac{1}{2}O_2(g) \longrightarrow CO(g)$ is 3.1% higher than the entropy change on $CO_2(g) \longrightarrow CO(g) + \frac{1}{2}O_2(g)$. Find the standard molar entropy of carbon monoxide.

Chapter H

Free Energy

H1 Entropy change of the surroundings

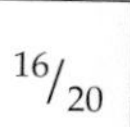

H1.1 If a reaction has an entropy change of the system of $-200\ \mathrm{J\,K^{-1}\,mol^{-1}}$, and an entropy change of the surroundings of $118\ \mathrm{J\,K^{-1}\,mol^{-1}}$, is it spontaneous?

H1.2 a) The enthalpy change of a system at constant pressure is $-24.0\ \mathrm{kJ\,mol^{-1}}$ at 298 K. Give the entropy change of the surroundings.

b) The entropy change of surroundings at constant pressure for a certain reaction at 540 K is $-120\ \mathrm{J\,K^{-1}\,mol^{-1}}$. Calculate the enthalpy change of the system.

c) If the entropy change of the surroundings for a process is $72.0\ \mathrm{J\,K^{-1}\,mol^{-1}}$, and the accompanying enthalpy change of the system is $-86.4\ \mathrm{kJ\,mol^{-1}}$, find the temperature.

d) The entropy change of the universe in a certain reaction is $45\ \mathrm{J\,K^{-1}\,mol^{-1}}$. If ΔS for the reaction is $-8.0\ \mathrm{J\,K^{-1}\,mol^{-1}}$, find the entropy change of the surroundings.

e) The total entropy change in the universe for a certain reaction is $90\ \mathrm{J\,K^{-1}\,mol^{-1}}$. If the entropy change of the system is $20\ \mathrm{J\,K^{-1}\,mol^{-1}}$, calculate the entropy change of the surroundings.

H1.3 a) A phase change at constant pressure has an enthalpy change of $58\ \mathrm{kJ\,mol^{-1}}$ and an entropy change of $85\ \mathrm{J\,K^{-1}\,mol^{-1}}$. Find the temperature at which the entropy change of the surroundings is equal in magnitude and opposite in sign to that of the system (i.e. the temperature at which the phase change occurs at this pressure.)

b) The neutralisation of one mole of acid at 305 K releases $180\ \mathrm{J\,K^{-1}}$of entropy into the surroundings at constant pressure. Find the enthalpy of neutralisation.

c) The reaction of 3.50 g of lithium raises the entropy of the surroundings by 220 $\mathrm{J\,K^{-1}}$ at constant pressure and 294 K. Find the molar enthalpy change for the reaction.

H1.4 a) At 650 K, a reaction has $\Delta G = -62\ \mathrm{kJ\,mol^{-1}}$. If the entropy change of the surroundings is 96 $\mathrm{J\,K^{-1}\,mol^{-1}}$, find ΔS for the reaction.

b) At 700 K the decomposition of calcium carbonate has an enthalpy change of 177.0 $\mathrm{kJ\,mol^{-1}}$. Find the entropy change of the surroundings under constant pressure at this temperature.

c) The entropy change of the surroundings during the reaction $N_2(g) + 3\,H_2(g) \rightleftharpoons 2\,NH_3(g)$ is 203 $\mathrm{J\,K^{-1}\,mol^{-1}}$ at 500 K. Find the enthalpy of formation of ammonia at this temperature.

d) At 298 K, the enthalpy of formation of NiS is $-82.0\ \mathrm{kJ\,mol^{-1}}$. Find the entropy change of the surroundings on forming 2.27 g of NiS from its elements at 298 K.

H1.5 a) At 450 K the decomposition of silver carbonate into the oxide and carbon dioxide has an entropy change of $-180\ \mathrm{J\,K^{-1}\,mol^{-1}}$ in the surroundings and 166 $\mathrm{J\,K^{-1}\,mol^{-1}}$ in the system. Find ΔG and K_p at this temperature.

b) At 298 K, the decomposition of HI into its elements has an enthalpy change of 4.8 $\mathrm{kJ\,mol^{-1}}$ and $K_p = 3.5 \times 10^{-2}$. Find the entropy changes of the surroundings and the system.

H1.6 The lattice enthalpy of RbF is $-778\ \mathrm{kJ\,mol^{-1}}$. Calculate the entropy change of the surroundings if 7.9 g of RbF is vaporised into its ions at 1000 K.

H1.7 The enthalpy of combustion of ethyne is $-1300\ \mathrm{kJ\,mol^{-1}}$. Calculate the entropy change of the surroundings when 1.00 mole of ethyne is burnt in excess oxygen at 2600 K.

H1.8 Combustion of methanol releases 726 $\mathrm{kJ\,mol^{-1}}$ of heat. If 0.010 moles of methanol react fully with oxygen at 850 °C, calculate the entropy change of the surroundings.

H1.9 Combustion of hexane releases 4163 kJ of heat per mole of hexane. Calculate the entropy change of the surroundings when 4.3 g of hexane is fully combusted at 600 °C.

H1.10 The specific heat capacity of water is 4180 $\mathrm{J\,kg^{-1}\,K^{-1}}$.The complete combustion of 0.92 g of methanoic acid raised the temperature of 0.40 kg of water from 18 °C to 21 °C. Calculate the entropy change of the surroundings at their ambient temperature of 18 °C.

H2 Free energy changes

9/11

H2.1 a) At 298 K the decomposition of hydrogen peroxide has an enthalpy change of −98 kJ mol^{-1} and an entropy change of 62.9 J K^{-1} mol^{-1}. Calculate the free energy change at constant pressure.

b) The free energy change for a reaction at constant pressure is −4.5 kJ mol^{-1}. If the temperature is 320 K and the entropy change is −12 J K^{-1} mol^{-1}, find the enthalpy change.

H2.2 The free energy change of formation of iodine(III) chloride is −22.3 kJ mol^{-1} at 298 K. Given that the enthalpy of formation is −89.5 kJ mol^{-1}, find the standard entropy change on formation of 1 mole of iodine(III) chloride from its elements in their reference states at 298 K.

H2.3 The standard enthalpy change on decomposition of magnesium carbonate is 100.6 kJ mol^{-1}, and the standard entropy change is 174.8 J K^{-1} mol^{-1}. Find the temperature at which its decomposition becomes spontaneous under standard conditions.

H2.4 The standard enthalpy of formation of copper(II) oxide at 290 °C is −157 kJ mol^{-1}. The standard entropy change for the same process is −41.9 J K^{-1} mol^{-1}. Find the standard Gibbs free energy change of formation of copper(II) oxide at this temperature.

H2.5 For a certain reaction at 600 K, the standard entropy change of the system is 26 J K^{-1} mol^{-1} and that of the surroundings is 114 J K^{-1} mol^{-1}. Find the accompanying standard Gibbs free energy change.

H2.6 At 1000 K, the reaction $H_2 + CO_2 \rightleftharpoons H_2O + CO$ has an equilibrium constant, K_p, of 0.955. Find the Gibbs free energy change at this temperature.

H2.7 The decomposition of calcium carbonate has an equilibrium constant of 0.13 when the standard Gibbs free energy change is equal to 18.0 kJ mol^{-1}. Find the temperature at which this occurs.

H2.8 The reaction $2\,SO_2 + O_2 \rightleftharpoons 2\,SO_3$ has an equilibrium constant of around 3.0×10^4 atm^{-1} at 975 °C. Find the accompanying change in Gibbs free energy.

H2.9 The displacement of hydrogen from acid by iron, $2\,H^+(aq) + Fe(s) \rightleftharpoons Fe^{2+}(aq) + H_2(g)$, has a standard cell potential of 0.44 V. Find the associated standard Gibbs free energy change. [See page viii for the Faraday constant.]

H2.10 The decomposition of silver bromide in light requires 1.00 eV of energy per silver atom produced. Calculate the standard free energy of formation of AgBr(s).

16/19

H3 Free energy cycles

Use the standard Gibbs free energy changes of formation given in the table to answer the questions that follow.

Substance	$\Delta_f G^\circ$ / kJ mol^{-1}	Substance	$\Delta_f G^\circ$ / kJ mol^{-1}
S(g)	238.3	$H_2O(l)$	−237.2
$S_2(g)$	79.3	$H_2O(g)$	−228.6
$SO_2(g)$	−300.2	HCl(g)	−95.2
$SO_3(l)$	−368.4	$CH_3COOH(l)$	−389.9
$SF_4(g)$	−731.4	$CH_3COCl(l)$	−208.0
$SF_6(g)$	−1105.4	$CH_3OH(l)$	−166.4
$S_2Cl_2(l)$	4.2	$CH_3Cl(g)$	−57.4
$SCl_2(g)$	−69.7	$CH_4(g)$	−50.8
$SOCl_2(l)$	−197.9	$CO_2(g)$	−394.4

H3.1 a) Find the standard free energy change for the hydrolysis of thionyl chloride, given by $SOCl_2(l) + H_2O(l) \longrightarrow SO_2(g) + 2\,HCl(g)$.

b) Find the standard free energy change for $SF_4(g) + F_2(g) \longrightarrow SF_6(g)$.

c) Find the standard free energy change for $S(g) + 3\,F_2(g) \longrightarrow SF_6(g)$.

d) Find the standard free energy change for $S_2Cl_2(l) + Cl_2(g) \longrightarrow 2\,SCl_2(g)$.

H3.2 Use the standard free energies of formation above to find the magnitude of the equilibrium constant of the contact process $2\,SO_2(g) + O_2(g) \rightleftharpoons 2\,SO_3(l)$, at 298 K. Use $R = 8.314$ J mol^{-1} K^{-1}.

H3.3 The oxidation of thionyl chloride by oxygen, $SOCl_2(l) + \frac{1}{2}O_2(g) \longrightarrow SO_2Cl_2(l)$, has a standard free energy change of −107.1 kJ mol^{-1}. Find the standard free energy of formation of $SO_2Cl_2(l)$.

H3.4 a) The standard free energy change for the gas phase reaction, $S_8(g) + 16\,F_2(g) \longrightarrow 8\,SF_4(g)$, is -5900.9 kJ mol^{-1}. Find the standard free energy of sublimation of sulfur, and hence its vapour pressure at 298 K, given that in units of atm, $K_p = p(S_8)$.

b) Use your answer to the previous part to find the standard free energy change and the equilibrium constant at 298 K for the reaction, $S_8(g) \rightleftharpoons 4\,S_2(g)$. Use $R = 8.314$ J mol^{-1} K^{-1}.

H3.5 Find the standard free energy of vaporization of water at 298 K, and hence its vapour pressure.

H3.6 a) Calculate the standard free energy change for the reaction, $CH_3COOH(l) + SOCl_2(l) \longrightarrow CH_3COCl(l) + SO_2(g) + HCl(g)$.

b) Calculate the standard free energy change for the hydrolysis of ethanoyl chloride, $CH_3COCl(l)$.

c) Add together the two free energy changes in parts (a) and (b) and subtract the free energy change for the hydrolysis of thionyl chloride. Give the total change in free energy.

H3.7 Two possible methods for the preparation of chloromethane are based on the reactions P and Q as shown.
Reaction P: $CH_3OH(l) + SOCl_2(l) \longrightarrow CH_3Cl(g) + SO_2(g) + HCl(g)$
Reaction Q: $CH_4(g) + Cl_2(g) \longrightarrow CH_3Cl(g) + HCl(g)$

a) Find the standard Gibbs free energy change for P

b) Find the standard Gibbs free energy change for Q

c) State which is likely to have the larger standard entropy change of the system.

d) State which is likely to have the larger standard entropy change of the surroundings.

H3.8 The isomerization of cyclopropane into propene has an equilibrium constant of 1.43×10^5 at 298 K.

a) Find the standard free energy change for the reaction.

b) Given that the standard free energy of formation of propene is 74.4 kJ mol^{-1}, find the standard free energy of formation of cyclopropane.

H3.9 a) Given that the standard free energy of formation of ethyne, $C_2H_2(g)$, is 209.2 kJ mol^{-1}, find its standard free energy of combustion.

b) Now use the standard free energy of combustion of benzene, −3202.5 kJ mol^{-1}, to calculate the standard free energy change for the reaction, $3\,C_2H_2(g) \longrightarrow C_6H_6(l)$.

5/7

H4 Thermodynamic stability

Free energy of formation at 298 K, in kJ mol^{-1}

FeO	−245.4	PbO	−187.7	Cu	0	$Cu(OH)_2$	−359.4
Fe_3O_4	−1015.5	Pb_3O_4	−601.2	Cu_2O	−146.0	H_2O	−237.2
Fe_2O_3	−742.2	PbO_2	−217.4	CuO	−129.7	NH_3	−16.5

H4.1 a) At 298 K, is copper(II) hydroxide stable with respect to decomposition?

b) Pb_3O_4 is a distinct compound rather than a mixture of PbO and PbO_2. Calculate the decrease in Gibbs free energy when the reaction between PbO and PbO_2 produces 1.000 mole of Pb_3O_4 at 298 K.

c) Hydrogen peroxide, H_2O_2, is thermodynamically unstable with respect to decomposition to water and oxygen at 298 K, but is stable with respect to its elements. In which range does the Gibbs free energy of formation of hydrogen peroxide lie? A) >237.2 kJ mol^{-1} B) > 0;<237.2 kJ mol^{-1} C) >−237.2 kJ mol^{-1};< 0 D) <−237.2 kJ mol^{-1}

H4.2 Iron, lead and copper all form a range of oxides, as shown in the table.

a) Identify the thermodynamically stable oxide of each element at 298 K in excess oxygen.

b) Identify the thermodynamically stable oxide of each element at 298 K with excess metal.

H4.3 a) Indicate which of the following could reduce copper(II) oxide to copper: H_2, NH_3, Fe, FeO, Fe_3O_4, Fe_2O_3, Pb, PbO, Pb_3O_4, Cu_2O

b) Indicate which of the following could oxidize copper(I) oxide to copper(II) oxide: Cu_2O, Fe_2O_3, PbO_2, H_2O, O_2

Chapter I

Equilibrium

I1 Equilibrium constant, K_p

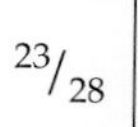

I1.1 Write expressions for K_p for the following gas phase equilibria:

a) $CH_3CHCH_2 \rightleftharpoons CH_2CH_2CH_2$

b) $3\,O_2 \rightleftharpoons 2\,O_3$

c) $H_2 + I_2 \rightleftharpoons 2\,HI$

d) $N_2 + 3\,H_2 \rightleftharpoons 2\,NH_3$

I1.2 Write balanced equations to show the equilibria represented by the following equilibrium constants, K_p.

a) $\dfrac{p(\mathrm{Cl})^2}{p(\mathrm{Cl_2})}$ b) $\dfrac{p(\mathrm{SO_3})^2}{p(\mathrm{SO_2})^2 p(\mathrm{O_2})}$ c) $\dfrac{p(\mathrm{CO})^2 p(\mathrm{H_2})^2}{p(\mathrm{CH_4}) p(\mathrm{CO_2})}$

I1.3 Complete the rows in the following table:

TOTAL PRESSURE	MOLE FRACTION	PARTIAL PRESSURE
1.0 atm	0.075	(a)
125 MPa	4.00×10^{-7}	(b)
4.0 lb ft^{-2}	0.30	(c)
50 bar	(d)	200 mbar
2.0 GPa	(e)	40 kPa
(f)	2.5×10^{-3}	1.4×10^4 Pa
(g)	80.0%	120 mmHg

I1.4 Express a mole fraction of 250.0 ppm as a percentage.

I1.5 The reaction

$$A + B \rightleftharpoons 2C$$

occurs in the gas phase. Its value of K_p at a temperature of 600 K is 2500. Each row in the table below shows possible partial pressures at equilibrium at 600 K. Find the missing value in each row.

$p(A)$	$p(B)$	$p(C)$
20,000 Pa	20,000 Pa	(a)
1.00 MPa	(b)	100 MPa
12.5 cm H_2O	3.75 cm H_2O	(c)
(d)	4.0×10^6 torr	1.60×10^8 torr
8.0×10^{-4} atm	5.0×10^{-4} atm	(e)

I1.6 The table below shows initial pressures before equilibration at 600 K for the reaction described in question 5. Fill in the rows by calculating the equilibrium pressures in each case.

	Initial $p(A)$	Initial $p(B)$	Initial $p(C)$	Equilib. $p(A)$	Equilib. $p(B)$	Equilib. $p(C)$
(a)	10.0 atm	10.0 atm	0.0 atm			
(b)	0.0 MPa	0.0 MPa	12.0 MPa			
(c)	250.0 bar	250.0 bar	0.0 bar			
(d)	0.00 psi	0.00 psi	2400.00 psi			

I1.7 At a temperature of 900 K, the equilibrium constant for the reaction in question 5 is $K_p = 850$. Is the reaction exothermic or endothermic in the forward direction?

I1.8 The reaction

$$CH_4 + H_2O \rightleftharpoons CO + 3\,H_2$$

has an equilibrium constant, K_p, of 150.5 Pa^2 at a temperature of 1073 K.[1]

$$K_p = \frac{p(CO)p(H_2)^3}{p(CH_4)p(H_2O)}$$

a) If the partial pressures at equilibrium are $p(CH_4) = 20$ kPa, $p(H_2O) =$ 20 kPa and $p(CO) = 50$ kPa, find the partial pressure of hydrogen at equilibrium.

b) If equal amounts of methane and steam are mixed and allowed to reach equilibrium, and the partial pressures $p(CO) = 40.0$ kPa and $p(H_2) = 120$ kPa, find the partial pressure of methane at equilibrium.

c) Once the gases have reached equilibrium, the total pressure is suddenly doubled by the engineer. Fill in the table below with 'increases', 'decreases' or 'stays the same' to describe what happens as the system reaches a new equilibrium:

K_p	
mole fraction of CO	
mole fraction of CH_4	
$p(H_2O)$	

I2 Equilibrium constant, K_c

34/42

I2.1 a) At equilibrium in the reaction $A(aq) + B(aq) \rightleftharpoons C(aq) + D(aq)$, the concentrations in mol dm^{-3} are: $[A]_{(eq)} = 0.25$, $[B]_{(eq)} = 0.10$, $[C]_{(eq)} = 0.030$ and $[D]_{(eq)} = 0.010$. Calculate K_c.

[1]Charles O. Hawk et al, Ind. Eng. Chem., 1932, **24**(1), pp. 23–27

b) The reaction $X(aq) + 3\,Y(aq) \rightleftharpoons 2\,Z(aq)$ reaches an equilibrium in which the three concentration in mol dm^{-3} are: $[X]_{(eq)} = 2.0 \times 10^{-4}$, $[Y]_{(eq)} = 1.6 \times 10^{-5}$, $[Z]_{(eq)} = 0.024$. Calculate the magnitude of K_c.

c) The reaction $A(aq) + B(aq) \rightleftharpoons C(aq) + H_2O(l)$ has an equilibrium constant equal to 0.050 dm^3 mol^{-1}. If the equilibrium concentration of A is 0.025 mol dm^{-3}, and that of B is 0.020 mol dm^{-3}, find the equilibrium concentration of C, in mol dm^{-3}.

d) K_c for the reaction $2\,J(aq) \rightleftharpoons K(aq) + L(aq)$ is found to be 28.2 and at equilibrium, there is 0.815 mol dm^{-3} of K and 1.24 mol dm^{-3} of L. Calculate the equilibrium concentration of J in mol dm^{-3} .

I2.2 The reaction $A(aq) \rightleftharpoons B(aq) + C(aq)$ has an equilibrium constant given by:

$$K_c = \frac{[B]_{(eq)}\,[C]_{(eq)}}{[A]_{(eq)}}$$

Where $[X]_{(eq)}$ is the equilibrium concentration of X in mol dm^{-3}.

a) Give the units of K_c.

b) If a 2.0 mol dm^{-3} solution of A is allowed to reach equilibrium, at which 1.2 mol dm^{-3} of A remains, find the equilibrium concentrations of B and C in mol dm^{-3}.

c) Find the value of K_c.

I2.3 The equilibrium constant for esterification can be generalised as

$$K_c = \frac{[\text{ester}]_{(eq)}\,[\text{water}]_{(eq)}}{[\text{alcohol}]_{(eq)}\,[\text{acid}]_{(eq)}}$$

when water is not the solvent. If the system reaches equilibrium and then more alcohol is added, will the concentration of acid increase, decrease or stay the same over the next few minutes?

I2.4 The units of K_c depend on the stoichiometry of the reaction, and on which species appear and are included in K_c. For each of the following processes, give the power to which concentration units (mol dm^{-3}) are raised to give the units of K_c, e.g. if K_c has units of mol^2 dm^{-6}, then the answer is 2.

a) $CH_3CHICHICOOH(aq) \rightleftharpoons CH_3CHCHCOOH(aq) + I_2(aq)$

b) $CH_3COCH_3(aq) \rightleftharpoons CH_3COHCH_2(aq)$

c) $Ag^+(aq) + 2\,NH_3(aq) \rightleftharpoons [Ag(NH_3)_2]^+(aq)$

d) $Cl_2(aq) + H_2O(l) \rightleftharpoons HCl(aq) + HOCl(aq)$

e) $2\,CrO_4^{2-}(aq) + 2\,H^+(aq) \rightleftharpoons Cr_2O_7^{2-}(aq) + H_2O(l)$

f) $SnCl_2(s) \rightleftharpoons Sn^{2+}(aq) + 2\,Cl^-(aq)$

g) $CaCO_3(s) + CO_2(aq) + H_2O(l) \rightleftharpoons Ca(HCO_3)_2(aq)$

h) $Hg_2^{2+}(aq) + 2\,Cl^-(aq) \rightleftharpoons Hg_2Cl_2(s)$

i) $3\,Br_2(aq) + 6\,NaOH(aq) \rightleftharpoons NaBrO_3(aq) + 5\,NaBr(aq) + 3\,H_2O(l)$

j) $ZnO(s) + H_2O(l) \rightleftharpoons Zn(OH)_2(aq)$

I2.5 The reaction, $2\,P(aq) + Q(aq) \rightleftharpoons R(aq) + S(aq)$ reaches equilibrium.

a) If equal volumes of R(aq) and S(aq), both with initial concentration 1.00 mol dm^{-3} are mixed and come to equilibrium at 320 K, the concentration, $[S]_{(eq)} = 0.422$ mol dm^{-3}. Find the equilibrium concentrations of P, Q and R in mol dm^{-3}, that is $[P]_{(eq)}$, $[Q]_{(eq)}$ and $[R]_{(eq)}$.

b) Find K_c.

c) If the reaction is exothermic, will the equilibrium constant be higher, lower, or the same at 330 K?

I2.6 For the reaction, $Fe^{2+}(aq)+Ag^+(aq) \rightleftharpoons Fe^{3+}(aq)+Ag(s)$, the equilibrium constant is given by:

$$K_c = \frac{[Fe^{3+}]_{(eq)}}{[Fe^{2+}]_{(eq)}\,[Ag^+]_{(eq)}}$$

In this case, $K_c = 1.17$ dm^3 mol^{-1} at RTP. Equal amounts of iron(II) sulfate and silver(I) nitrate in solution were mixed and left to stand until equilibrium was reached and a solution was obtained that contained iron(III) at a concentration of 6.2×10^{-4} mol dm^{-3}.

a) Calculate $[Ag^+]_{(eq)}$ in mol dm^{-3}.

b) Calculate the concentration of iron(II) sulfate in the solution immediately after mixing, before any reaction occurred, in mol dm^{-3}.

I2.7 The reaction $F(aq) \rightleftharpoons G(aq)$ has an equilibrium constant of 4.0. State the equilibrium constants of the following reactions under the same conditions:

a) $G(aq) \rightleftharpoons F(aq)$

b) $2\,F(aq) \rightleftharpoons 2\,G(aq)$

c) $2\,F(aq) \rightleftharpoons F(aq) + G(aq)$

I2.8 The reaction $A(aq) + B(aq) \rightleftharpoons C(aq)$ reaches equilibrium in 150 cm^3 of water. If 25 cm^3 of water slowly evaporates, so that equilibrium is maintained, will the concentration of A increase, decrease, or stay the same?

I2.9 The hydration of chloroethanal has an equlibrium constant, K_c, of 37.0.[2]

$$ClCH_2CHO(aq) + H_2O(l) \rightleftharpoons ClCH_2CH(OH)_2(aq).$$

Complete the following table by providing any missing initial or equilibrium concentrations:

Initial $[ClCH_2CHO]$	Initial $[ClCH_2CH(OH)_2]$	Equilibrium $[ClCH_2CHO]$	Equilibrium $[ClCH_2CH(OH)_2]$
1.20 $mol\,dm^{-3}$	0.00 $mol\,dm^{-3}$	(a) $mol\,dm^{-3}$	(b) $mol\,dm^{-3}$
0.00 $mol\,dm^{-3}$	(c) $mol\,dm^{-3}$	(d) $mol\,dm^{-3}$	0.292 $mol\,dm^{-3}$
(e) $mol\,dm^{-3}$	0.100 $mol\,dm^{-3}$	0.0184 $mol\,dm^{-3}$	(f) $mol\,dm^{-3}$
14.0 $mmol\,dm^{-3}$	0.800 $mmol\,dm^{-3}$	(g) $mmol\,dm^{-3}$	(h) $mmol\,dm^{-3}$
161 $mg\,dm^{-3}$	0.00 $mg\,dm^{-3}$	(i) $mg\,dm^{-3}$	(j) $mg\,dm^{-3}$

k) Trichloroethanal is almost fully hydrated in aqueous solution, with a K_c value[3] of around 10^4, to give "chloral hydrate". Give the approximate concentration in $mol\,dm^{-3}$ of chloral hydrate, to 2 s.f., required to maintain an equilibrium concentration of 1.0 $nmol\,cm^{-3}$ of the free, unhydrated form.

[2] Tadashi Okuyama, Howard Maskill, 'Organic Chemistry: A Mechanistic Approach', OUP Oxford, 2013; ISBN 9780199693276.

[3] *Ibid.*

I2.10 Magnesium oxide is partially hydrolysed in water:
$MgO(s) + H_2O(l) \rightleftharpoons Mg(OH)_2(aq)$. The equilibrium constant, K_c = 2.0×10^{-4} mol dm^{-3} for this reaction.

a) Give the concentration of a saturated solution of magnesium hydroxide.

b) Give the amount of magnesium hydroxide that is present in 95 cm^3 of water when 0.5 g of magnesium oxide powder is added to it.

c) If the volume of water is then doubled does the concentration of magnesium hydroxide at equilibrium increase, decrease or stay the same?

d) In the doubled volume of water, does the mass of magnesium oxide remaining increase, decrease or stay the same?

I3 Solubility product

15/18

All questions refer to a temperature of 298 K, unless otherwise stated. The density of water may be taken to be 1.00 g cm^{-3} at RTP.

I3.1 a) The solubility product of copper(II) chromate(VI), $CuCrO_4$, at 298 K is 3.6×10^{-6} mol^2 dm^{-6}. Calculate the concentration of a saturated solution of copper(II) chromate(VI) at this temperature.

b) The solubility product of silver(I) bromate(V), $AgBrO_3$, is 6.0×10^{-5} mol^2 dm^{-6}. Calculate the concentration of a saturated solution in g dm^{-3}.

I3.2 a) A saturated solution of barium sulfate, $BaSO_4$, has a concentration of 1.0×10^{-5} mol dm^{-3}. Calculate its solubility product.

b) The solubility of barium carbonate, $BaCO_3$, is given as 9.12×10^{-6} mol/100g. Calculate its solubility product in mol^2 dm^{-6}.

I3.3 a) Does the solubility product of a solid typically increase, decrease or stay the same on an increase in temperature?

b) Is silver(I) chloride more soluble or less soluble in a solution of sodium chloride than it is in deionized water?

I3.4 a) The solubility product of silver(I) chloride is 2.0×10^{-10} mol^2 dm^{-6}. Calculate the concentration of silver ions present in solution when excess solid silver(I) chloride is added to sodium chloride solution with a concentration of 1.25×10^{-3} mol dm^{-3} and allowed to reach equilibrium.

b) Calculate the mass of silver(I) chloride that will dissolve in 1.0 dm^3 of water.

I3.5 The solubility product of calcium sulfate is 2.0×10^{-5} mol^2 dm^{-6}. Calculate the mass of calcium sulfate that will dissolve in 250 cm^3 of water.

I3.6 The solubility product of mercury(II) selenide is 1.0×10^{-59} mol^2 dm^{-6}.

a) Calculate the volume of water required in theory to dissolve a single HgSe unit.

b) Calculate the volume of water required to dissolve 1.0 mole of HgSe.

I3.7 The solubility products of barium oxalate, BaC_2O_4, and iron(II) oxalate, FeC_2O_4, are 2.1×10^{-7} and 1.7×10^{-7} mol^2 dm^{-6} respectively. If a small amount of water is added to a 1 : 1 mixture of the two solids and a solution in equilibrium is formed, find the concentrations of oxalate, iron(II) and barium ions in the solution:

a) $[C_2O_4^{2-}]_{eq}$

b) $[Fe^{2+}]_{eq}$

c) $[Ba^{2+}]_{eq}$

I3.8 The solubility product of lead(II) chloride is 2.0×10^{-5} mol^3 dm^{-9}. Calculate the concentration of a saturated solution of lead(II) chloride.

I3.9 The solubility product of lead(II) arsenate(V), $Pb_3(AsO_4)_2$, is 4.1×10^{-36} mol^5 dm^{-15}. If a solution is formed in equilibrium with the solid, calculate the concentrations of:

a) lead(II) arsenate(V)

b) lead(II)

c) arsenate(V)

I3.10 60.0 cm^3 of a solution containing thallium(I) ions was treated with 20.0 cm^3 of aqueous sodium chloride with a concentration of 1.00 mol dm^{-3}. The precipitated thallium(I) chloride was weighed and found to have a mass of 0.173 g. The solubility product of thallium(I) chloride is 1.75×10^{-4} mol^2 dm^{-6}. Calculate the concentration of thallium(I) ions in the original solution before treatment.

I4 Partition

I4.1 A solute, S, is dissolved in two immiscible liquids, water and toluene. After reaching equilibrium, [S] in water is found to be 0.480 mol dm^{-3}, and [S] in toluene is found to be 0.060 mol dm^{-3}. Calculate the partition coefficient of S between water and toluene.

I4.2 A solute, S, has a partition coefficient of 0.080 between two solvents. If [S] in the first solvent is 0.032 mol dm^{-3}, calculate [S] in the second solvent.

I4.3 10.0 mg of solid is dissolved in water and then shaken with an equal volume of the immiscible solvent, 2-methylbutan-2-ol. If 9.0 mg of solid ends up in the 2-methylbutan-2-ol, calculate the partition coefficient of solid between water and 2-methylbutan-2-ol.

I4.4 40 cm^3 of water containing 0.060 g of a solute is shaken with 10 cm^3 of ether. Once the two phases separate from each other, the water is removed and found to contain only 0.0020 g of solute. Assuming that equilibrium was established, calculate the water-ether partition coefficient for this solute.

I4.5 Fill in the missing values in the table:

Solvent A volume	Solvent B volume	Mass of solute in solvent A	Mass of solute in solvent B	Partition constant $K_{A/B}$
120 cm^3	80 cm^3	0.60 g	4.0 mg	(a)
35 ml	25 ml	2.0 mg	(b)	0.026
6.0 fl oz	(c)	0.030 oz	0.010 oz	8.0
1.20 dm^3	40.0 cm^3	(d)	360 μg	6750
160 dm^3	20.0 m^3	120 g	(e)	2.81×10^7

I4.6 5.0 g of solute Y is dissolved in solvent A to give a solution which is then shaken with an equal volume of solvent B. If the partition coefficient for Y between A and B is 6.2, calculate the mass of Y dissolved in each solvent at equilibrium.

I4.7 A solution of an organic compound, X, in 125.0 cm^3 of water is shaken with 50.00 cm^3 of ether and the layers allowed to separate. The water-ether partition coefficient of X is 1.820×10^{-4}. Equilibrium is reached and the water still contains 0.1400 mg of X. Before the layers are separated, however, 30.00 cm^3 of the ether slowly evaporates while equilibrium is maintained. Calculate the mass of X remaining in the ether after separation.

I4.8 The octanol-water partition coefficient of benzaldehyde is given as 30.2.[4] A solution of benzaldehyde in 50.0 cm^3 of water is shaken with 30.0 cm^3 of octanol until equilibrium is established. The octanol is isolated and a second portion of 30.0 cm^3 of octanol is used to repeat the process on the 50.0 cm^3 of aqueous solution that remains. The two octanol portions are then combined to give a single solution of benzaldehyde in octanol.

a) Find the percentage of the benzaldehyde extracted into the first portion of octanol.

b) Find the percentage of the total mass of benzaldehyde that is removed into the second portion of octanol.

c) Give the final concentration of benzaldehyde in the combined octanol portions if the initial concentration in water was 0.0412 mol dm^{-3}.

I4.9 The partition coefficient between water and octanol of butanethiol is 5.25×10^{-3}.[5] Calculate the volume of octanol required to remove 99.0 % of any butanethiol dissolved in 75.0 cm^3 of water.

I4.10 40.0 cm^3 of water contains 1.25 g of solute. If an immiscible solvent has a partition coefficient for this solute of 17.1 against water, calculate the volume of this solvent required to remove 1.20 g of the solute from the water.

---

[4] James Sangster, 'Octanol-Water Partition Constants of Simple Organic Compounds', JPCRD **18**(3) 1989 pp. 1111–1227

[5] *Ibid.*

I5 $RT \ln K$

$^{15}/_{18}$

I5.1 Give the value to 3 s.f., and units of the gradient of a graph of $\Delta G^\circ/T$ against $\ln K$.

I5.2 Complete the table by filling in the missing value in each row [use $R = 8.314$ J mol^{-1} K^{-1}]:

ΔG°	T	K_p
(a)	298 K	10.0
−15.0 kJ mol^{-1}	298 K	(b)
1.62 kJ mol^{-1}	(c)	0.54
(d)	40 °C	1.08
−118 kJ mol^{-1}	550 °C	(e)
−23.0 kJ mol^{-1}	(f)	1210
(g)	292 K	1.22×10^{-8}
31.0 kJ mol^{-1}	224 K	(h)
−888 J mol^{-1}	(i)	4.01

I5.3 $\ln K_p$ for a certain reaction increases by a factor of 1.9 when the temperature doubles from 300 K to 600 K. Find the factor by which the magnitude of the (negative) ΔG° increases.

I5.4 A change in temperature from 300 K to 400 K changes ΔG° for a reaction from 1.8 kJ mol^{-1} to −3.1 kJ mol^{-1}. Find the factor by which K_p increases.

I5.5 A decrease in temperature from 90 °C to 50 °C causes ΔG° to increase in magnitude by a factor of 1.62. Find the percentage increase in $\ln K$.

I5.6 The reaction of sulfur dioxide with oxygen in the gas phase proceeds as $2\,SO_2(g) + O_2(g) \rightleftharpoons 2\,SO_3(g)$. If the partial pressure of sulfur dioxide at equilibrium at 500 K is 6.8×10^{-4} atm, that of oxygen is 4.6×10^{-4} atm, and that of sulfur trioxide is 2.3 atm, find ΔG° at this temperature.

I5.7 The important industrial reaction, $H_2(g) + CO_2(g) \rightleftharpoons CO(g) + H_2O(g)$ reaches equilibrium at 427 °C when the partial pressures are: $p(H_2) =$ 135 atm; $p(CO_2) = 130$ atm; $p(CO) = 52.0$ atm; $p(H_2O) = 41.5$ atm. Find ΔG° at this temperature.

I5.8 The reaction $A(g) + 2\,B(g) \rightleftharpoons C(g) + D(g)$ has a standard Gibbs free energy change of 620 J mol^{-1} at 250 °C. If $p(C)$ and $p(D)$ at equilibrium are 2.6 atm, and $p(A)$ at equilibrium is 0.81 atm, find the equilibrium partial pressure of B.

I5.9 The dissociation of dinitrogen tetroxide, $N_2O_4(g) \rightleftharpoons 2\,NO_2(g)$ at 400 K and a constant pressure of 73.6 atm, produces mole fractions of 98.5% NO_2 and 1.5% N_2O_4 at equilibrium. Find ΔG° at this temperature.

I5.10 The direct combination of nitrogen and oxygen, $N_2(g) + O_2(g) \rightleftharpoons 2\,NO(g)$, has $\Delta G^{\circ} = 144$ kJ mol^{-1} at 1500 K. Find the equilibrium partial pressure of NO if the equilibrium partial pressures of N_2 and O_2 are 92 atm and 25 atm respectively.

Chapter J

Acids & Bases

J1 Brønsted-Lowry & Lewis

$^{31}/_{38}$

J1.1 Identify the following compounds as Brønsted-Lowry acids or bases in aqueous solution:

a) NH_3
b) $NaHCO_3$
c) HF
d) CH_3COOH
e) H_3PO_4
f) HCN
g) $NaHSO_4$
h) C_6H_5OH
i) $HCOOK$
j) N_2H_4
k) NH_4Cl
l) Na_3PO_4

J1.2 Identify the following as Lewis acids or bases:

a) Br^-
b) Cu^{2+}
c) H^+
d) H^-
e) CH_3NH_2
f) SbF_5
g) CN^-
h) C_5H_5N
i) C_2H_4

J1.3 Identify the acid-base reactions in the list below:

a) $HNO_3 + KOH \longrightarrow KNO_3 + H_2O$

b) $Mg(OH)_2 + 2\,HCl \longrightarrow MgCl_2 + 2\,H_2O$

c) $Zn + CuSO_4 \longrightarrow ZnSO_4 + Cu$

d) $C_2H_4O + H_2O_2 \longrightarrow CH_3COOH + H_2O$

e) $CaO + SiO_2 \longrightarrow CaSiO_3$

f) $2\,NH_3 + H_2SO_4 \longrightarrow (NH_4)_2SO_4$

g) $Fe + 2\,HCl \longrightarrow FeCl_2 + H_2$

h) $ZnCO_3 + 2\,HNO_3 \longrightarrow Zn(NO_3)_2 + H_2O + CO_2$

i) $SO_3 + H_2S_2O_7 \longrightarrow 2\,H_2SO_4$

j) $AlCl_3 + NaCl \longrightarrow NaAlCl_4$

J1.4 Give the conjugate acid of NH_3.

J1.5 Give the conjugate base of H_2SO_4.

J1.6 Give the conjugate base of NH_3.

J1.7 Give the conjugate base of $H_3N^+CH_2COO^-$.

J1.8 Give the conjugate acid of $H_3N^+CH_2COO^-$.

J1.9 Give the conjugate acid of PO_4^{3-}.

J1.10 Give the conjugate base of CH_3OH.

Additional Questions

J1.11 Indicate the proton (H^+) donated most readily by each of the following Brønsted-Lowry acids. Where more than one correct proton is equivalent, indicate any one of the equivalent protons.

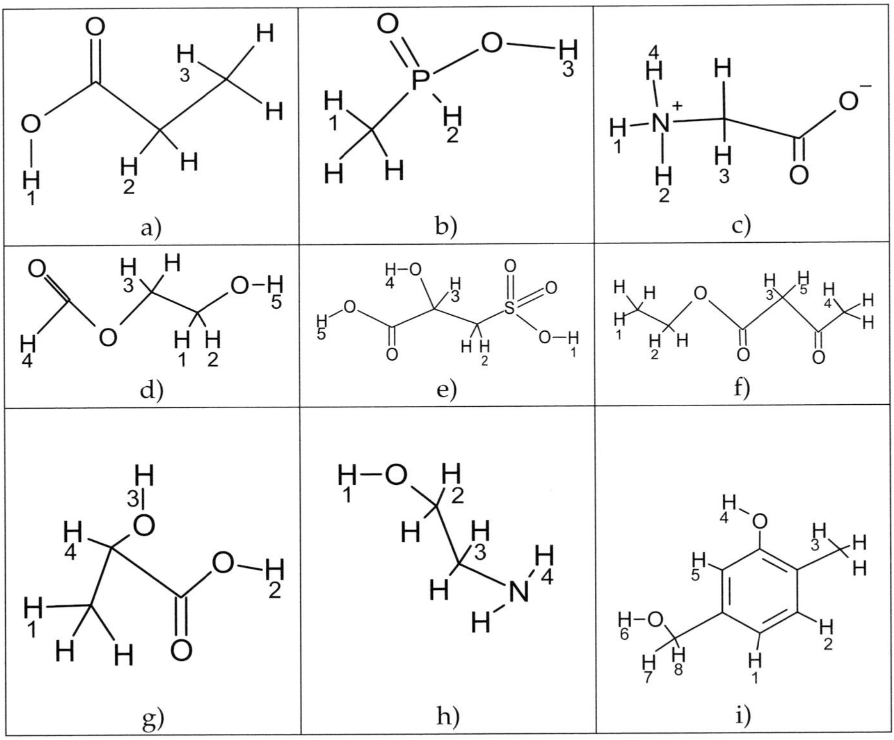

a)	b)	c)
d)	e)	f)
g)	h)	i)

J1.12 Indicate the atom which most readily donates its lone pair of electrons on the following Lewis bases.

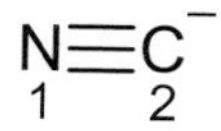 a)	 b)	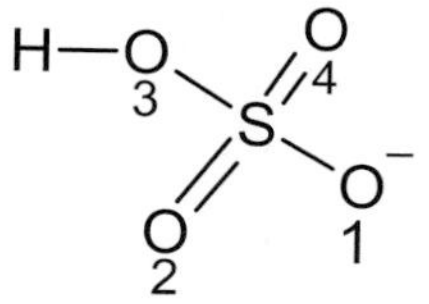 c)
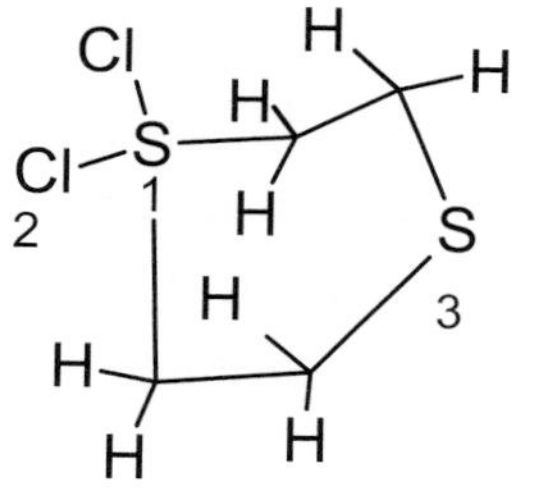 d)	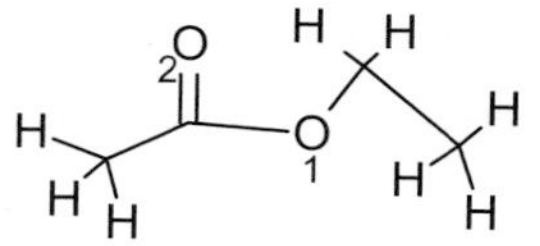 e)	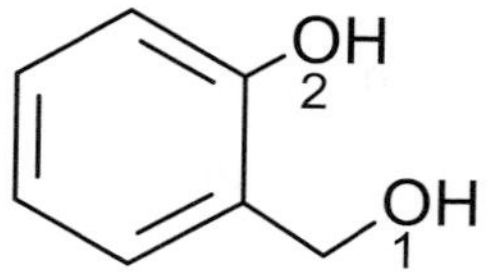 f)
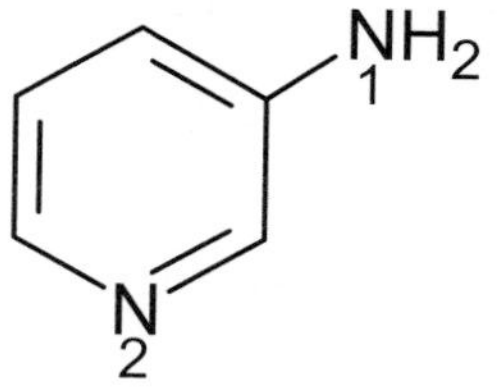 g)	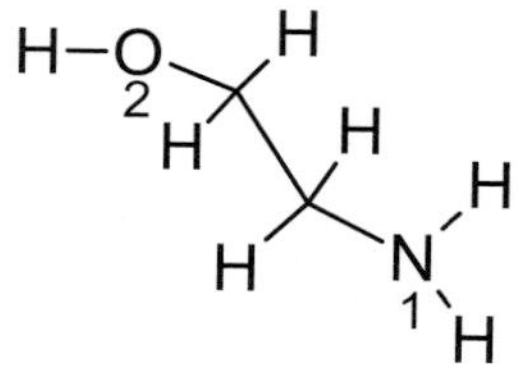 h)	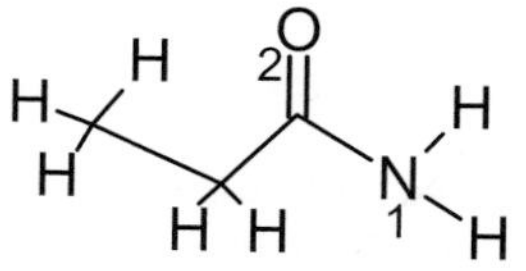 i)

J2 pH & K_w

17/21

J2.1 a) Give the units of K_w.

b) Give the value of K_w, in the usual units.

J2.2 a) Give the pH of pure water at 298 K.

b) Calculate the pH of a solution containing 2.5 mol dm^{-3} H^+.

c) Calculate the pH of a solution containing 1.8×10^{-4} mol dm^{-3} H^+.

d) Calculate the pH of a solution containing 0.90 mmol dm^{-3} H^+.

e) Calculate the H^+ ion concentration in a solution with pH = 5.2.

J2.3 a) Calculate the pH of an aqueous solution at 298 K containing 3.6×10^{-4} mol dm^{-3} of OH^-.

b) Calculate the OH^- ion concentration, in mol dm^{-3}, in an aqueous solution at 298 K and with a pH of 11.2.

c) Calculate the OH^- ion concentration, in mol dm^{-3}, in an aqueous solution at 298 K and with a pH of 4.90.

J2.4 a) Calculate the pH of a 0.012 mol dm^{-3} solution of HCl.

b) Calculate the pH of a 0.030 mol dm^{-3} solution of sulfuric acid.

c) Calculate the concentration of a solution of nitric acid with pH 2.1.

d) Calculate the concentration of a solution of sulfuric acid with a pH of 4.7.

J2.5 a) Calculate the concentration of a solution of barium hydroxide with a pH of 9.5.

b) Calculate the pH of a 0.0800 mol dm^{-3} solution of KOH.

J2.6 4.0 dm^3 of hydrogen chloride gas at RTP is dissolved in 2.2 dm^3 of water. Calculate the pH of the resulting solution.

J2.7 14 g of sulfuric acid is dissolved in 500 m^3 of water. Calculate the pH of the resulting solution.

J2.8 100 cm^3 of a solution of 0.750 mol dm^{-3} sulfuric acid is mixed with 400 cm^3 of a solution of 0.300 mol dm^{-3} sodium hydroxide. Calculate the pH of the resulting mixture.

J2.9 50 cm^3 of a solution of 0.200 mol dm^{-3} nitric acid is mixed with 200 cm^3 of a solution of 0.160 mol dm^{-3} potassium hydroxide. Calculate the pH of the resulting mixture.

J2.10 Pure water at 45 °C has a K_w of 4.0×10^{-14}. Give the pH of neutral water at 45 °C.

J3 K_a & pK_a

J3.1 Calculate the pH of a 0.050 mol dm^{-3} solution of iron(III) chloride.

Species	K_a/ mol dm^{-3}
Benzoic acid	6.3×10^{-5}
Hydrogen sulfide	8.9×10^{-8}
Iron(III)	6.0×10^{-3}
Methanoic acid	1.6×10^{-4}
Sulfuric(IV) acid	1.5×10^{-2}
Boric acid	5.8×10^{-10}

J3.2 Calculate the concentration of a solution of benzoic acid with a pH of 3.2.

J3.3 Give the pK_a of hydrogen sulfide.

J3.4 The pK_a of ethanoic acid is 4.8. Calculate its K_a.

J3.5 Calculate the pH of a 16 mmol dm^{-3} solution of boric acid.

J3.6 Calculate the K_a and pK_a of an acid, HA, with a pH of 5.0 when its concentration is 0.20 mol dm^{-3}.

J3.7 a) Calculate the pH of a 13.8 g dm^{-3} solution of methanoic acid.

b) 240 cm^3 of hydrogen sulfide gas is dissolved 500 cm^3 of water. Calculate the pH of the resulting solution.

J3.8 Calculate the approximate concentration of H^+ ions in a 0.12 mol dm^{-3} solution of iron(III) nitrate.

J3.9 Sulfur(IV) oxide dissolves in water to give sulfuric(IV) acid: $SO_2(g) + H_2O(l) \rightleftharpoons H_2SO_3(aq)$. Calculate the RTP volume of sulfur(IV) oxide required to reduce the pH of a lake of volume 0.40 km^3 from 7.0 to 6.0.

J3.10 An ant bite injects 150 ng of methanoic acid into a neutral aqueous region of volume 1 mm^3. Find the new pH in the region of the bite.

8/10

J4 K_b & pK_b

Species	K_b/ mol dm^{-3}
Lead(II) hydroxide	9.6×10^{-4}
Zinc hydroxide	9.6×10^{-4}
Ammonia	1.8×10^{-5}
Hydroxylamine	1.1×10^{-8}
Sulfide	7.9×10^{-2}
Cyanide	2.0×10^{-5}

J4.1 Calculate the pK_b of lead(II) hydroxide

J4.2 Pyridine has a pK_b value of 8.75. Calculate its K_b.

J4.3 Calculate the concentration of a solution of ammonia with pH 11.0.

J4.4 Calculate the pOH of a 0.15 mol dm^{-3} solution of hydroxylamine.

J4.5 Give the pK_a of $^{+}NH_3OH$ at 298 K.

J4.6 Give the K_a of hydrogen cyanide at 298 K.

J4.7 Calculate the mass of zinc in 200 cm^3 of zinc hydroxide solution at pH 11.0.

J4.8 Calculate the concentration of H^+ ions in 5.0 mol dm^{-3} ammonia solution.

J4.9 Calculate the mass of sodium cyanide required to give 50 cm^3 of an aqueous solution of pH 10.6.

J4.10 A solution of potassium sulfide in 100 cm^3 of water has a pH of 11.9. When treated with mercury(II) chloride, all the sulfide is precipitated as mercury(II) sulfide. Calculate the mass of mercury(II) sulfide obtained.

10/12

J5 Buffers

J5.1 Sodium dihydrogenphosphate has a pK_a value of 7.2. Give the pH of a buffer formed by mixing equal amounts of sodium dihydrogenphosphate and disodium hydrogenphosphate in aqueous solution.

J5.2 Propanoic acid has a pK_a value of 4.9 and is highly soluble in water. If 200 cm^3 of propanoic acid solution at 2.0 mol dm^{-3} is treated with 800 cm^3 of potassium propanoate solution at 1.0 mol dm^{-3}, give the pH of the resulting buffer.

J5.3 Given that benzoic acid has a K_a of 6.3×10^{-5} mol dm^{-3}, calculate the pH of a buffer containing equal amounts of benzoic acid and sodium benzoate.

J5.4 Given that methanoic acid has a K_a of 1.6×10^{-4} mol dm^{-3},

a) calculate the pH of a solution containing 25 mmol of methanoic acid and 40 mmol of potassium methanoate.

b) calculate the pH of a solution containing 0.40 mol of methanoic acid and 0.32 mol of magnesium methanoate.

J5.5 Given that methanoic acid has a K_a of 1.6×10^{-4} mol dm^{-3},

a) calculate the pH obtained when 100 cm^3 of 0.25 mol dm^{-3} methanoic acid is treated with 10 cm^3 of 0.50 mol dm^{-3} sodium hydroxide.

b) calculate the pH of the solution obtained when 1.7 g of sodium methanoate is dissolved in 40 cm^3 of 0.10 mol dm^{-3} hydrochloric acid.

J5.6 Calculate the volume of 2.00 mol dm^{-3} KOH that should be added to 60.0 cm^3 of 1.00 mol dm^{-3} H_3PO_4 to make a buffer solution of pH 2.00, given the pK_a of phosphoric(V) acid is 2.1. [Hint: work out the quantity in moles of acid used and then alkali required, rather than trying to use concentrations throughout.]

J5.7 Calculate the mass of sodium carbonate decahydrate, $Na_2CO_3 \cdot 10\,H_2O$, that should be added to 2.5 dm^3 of 0.40 mol dm^{-3} nitric acid to make a buffer of pH 10.5, given that the pK_a of hydrogencarbonate is 10.3. [Assume: (1) That no CO_2 is given off in the reaction. (2) That the nitric acid just determines the hydrogencarbonate concentration and does not participate in the buffer, so that the hydrogencarbonate concentration is obtained from the reaction, $HNO_3 + Na_2CO_3 \longrightarrow NaHCO_3 + NaNO_3$]

J5.8 A buffer solution made from "CHES" has a pH of 8.8 and contains 300 μmol of CHES and 95 μmol of its conjugate base. Calculate the pK_a and K_a of CHES.

J5.9 A buffer made from "hexamine" containing 0.00250 mol dm^{-3} of hexamine and 0.00180 mol dm^{-3} of its conjugate acid has a pH of 5.03. Calculate the pK_b of hexamine.

J5.10 A buffer of pH 7.8 is prepared by taking 200 cm^3 of 0.020 mol dm^{-3} "tris" solution and adding dilute hydrochloric acid from a burette until the pH is correct. If this requires 1.35 cm^3 of 2.0 mol dm^{-3} HCl(aq), calculate the pK_a of "tris".

Chapter K

Redox

$^{33}/_{41}$

K1 Oxidation number

K1.1 Give the oxidation number of nitrogen in the following compounds:

a) NH_3
b) NO
c) N_2
d) NO_2
e) HNO_3
f) $Ca(NO_3)_2$
g) N_2H_4
h) Mg_3N_2
i) NCl_3
j) NO^+

K1.2 Write down the oxidation number of:

a) Oxygen in H_2O
b) Sulfur in H_2SO_4
c) Phosphorus in H_3PO_4
d) Phosphorus in H_3PO_3
e) Chlorine in ClO_2
f) Oxygen in OF_2
g) Nitrogen in sodium nitrite
h) Nitrogen in ammonium sulfate
i) Oxygen in hydrogen peroxide
j) V in VO_2^+

k) V in VO^{2+}

l) Hg in Hg_2^{2+}

m) Cr in $Cr_2O_7^{2-}$

n) Mn in MnO_4^-

o) I in I_3^-

K1.3 Underline the species that is being reduced.

a) $CuO + H_2 \longrightarrow Cu + H_2O$

b) $C_3H_6 + H_2 \longrightarrow C_3H_8$

c) $2\,Na + Br_2 \longrightarrow 2\,NaBr$

d) $H_2O_2 + 2\,FeSO_4 + H_2SO_4 \longrightarrow 2\,H_2O + Fe_2(SO_4)_3$

e) $ZnCl_2 \longrightarrow Zn + Cl_2$

f) $Fe_2(SO_4)_3 + Zn \longrightarrow 2\,FeSO_4 + ZnSO_4$

g) $NiSO_4 + Fe \longrightarrow FeSO_4 + Ni$

h) $4\,C_3H_6O + NaBH_4 + 4\,H_2O \longrightarrow 4\,C_3H_8O + NaB(OH)_4$

K1.4 Underline the species that is being oxidized.

a) $2\,Al + Cr_2O_3 \longrightarrow Al_2O_3 + 2\,Cr$

b) $2\,NH_3 + 3\,CuO \longrightarrow N_2 + 3\,Cu + 3\,H_2O$

c) $2\,Cu^{2+} + 4\,I^- \longrightarrow 2\,CuI + I_2$

d) $6\,PbO + O_2 \longrightarrow 2\,Pb_3O_4$

e) $H_2O_2 + SO_2 \longrightarrow H_2SO_4$

f) $3\,H_2SO_4 + 2\,NaBr \longrightarrow 2\,NaHSO_4 + Br_2 + SO_2 + 2\,H_2O$

g) $Mg + 2\,CH_3COOH \longrightarrow Mg(CH_3COO)_2 + H_2$

h) $2\,Fe^{3+} + 6\,ClO^- + 4\,OH^- \longrightarrow 2\,FeO_4^{2-} + 3\,Cl_2 + 2\,H_2O$

K2 Half-equations

16/20

K2.1 Half-equations to complete and balance (assume all are in aqueous solution):

a) $Fe^{2+} \longrightarrow Fe^{3+}$ ____________________

b) Cu^{2+} ____________________ $\longrightarrow Cu$

c) MnO_4^- ____________________ $\longrightarrow Mn^{2+}$ ____________________

d) H_2O_2 ________________ $\longrightarrow 2\,OH^-$

e) $H_2O_2 \longrightarrow O_2 + 2\,H^+$ ________________

f) NH_3 ________________ $\longrightarrow NO +$ ________________

g) $Br^- +$ ________________ $\longrightarrow BrO_3^-$ ________________

h) $?\,NO_3^-$ ________________ $\longrightarrow N_2 +$ ________________

K2.2 Write reduction half-equations to show:

a) Reduction of iron(III) to iron metal

b) Reduction of bromine to bromide

c) Reduction of hydrogen ions to hydrogen

d) Reduction of water to hydrogen

e) Reduction of sulfate to sulfite

f) Reduction of chlorate(V) to chlorine

K2.3 Write oxidation half-equations to show:

a) Oxidation of cobalt(II) to cobalt(III)

b) Oxidation of iodide to iodine

c) Oxidation of bromine to bromate(V)

d) Oxidation of water to oxygen

e) Oxidation of hydrogen peroxide to oxygen

f) Oxidation of thiosulfate to tetrathionite ($S_4O_6^{2-}$)

16/20

K3 Balancing redox equations

K3.1 Balance the following redox equations:

a) $H_2(g) + ?Ag^+(aq) \longrightarrow ?H^+(aq) + ?Ag(s)$

b) $?Fe(s) + ?Cl_2(g) \longrightarrow ?FeCl_3(s)$

c) $I_2(aq) + ?S_2O_3^{2-}(aq) \longrightarrow ?I^-(aq) + S_4O_6^{2-}(aq)$

d) $CH_4(g) + ?Cl_2(g) \longrightarrow CCl_4(g) + ?HCl(g)$

e) $C_6H_{12}O_6(aq) + ?O_2(g) \longrightarrow ?CO_2(g) + ?H_2O(l)$

f) $?K(s) + ?H_2O(l) \longrightarrow ?KOH(aq) + H_2(g)$

g) $?Co^{3+}(aq) + ?H_2O(l) \longrightarrow ?Co^{2+}(aq) + ?H^+(aq) + O_2(g)$

h) $?Fe^{2+}(aq) + MnO_4^-(aq) + ?H^+(aq) \longrightarrow ?Fe^{3+}(aq) + Mn^{2+}(aq) + ?H_2O(l)$

i) $?Zn(s) + ?VO_2^+(aq) + ?H^+(aq) \longrightarrow ?Zn^{2+}(aq) + ?V^{2+}(aq) + ?H_2O(l)$

j) $C_2O_4H_2(aq) + H_2O_2(aq) \longrightarrow ?CO_2(g) + ?H_2O(l)$

k) $?BaFeO_4(s) + ?HCl(aq) \longrightarrow ?BaCl_2(aq) + ?FeCl_3(aq) + ?H_2O(l) + ?Cl_2(g)$

l) $?CH_3CH_2CH_2OH(l) + Cr_2O_7^{2-}(aq) + ?H^+(aq) \longrightarrow ?CH_3CH_2CHO(l) + ?Cr^{3+}(aq) + ?H_2O(l)$

K3.2 Complete the balanced equations to show the reactions between the following pairs of substances in acidic aqueous conditions (no fractions).

a) manganate(VII) and hydrogen peroxide

$$5\,H_2O_2(aq) + 2\,MnO_4^-(aq) + \_\_\_\_\_ \longrightarrow 2\,Mn^{2+}(aq) + 8\,H_2O(l) + \_\_\_\_\_$$

b) scandium and ethanoic acid

$$2\,Sc(s) + ?CH_3COOH(aq) \longrightarrow \_\_\_\_\_ + 3\,H_2(g)$$

c) ethanol and boiling dichromate(VI)

$$3\,CH_3CH_2OH(aq) + 2\,Cr_2O_7^{2-}(aq) + ?H^+(aq) \longrightarrow 3\,CH_3COOH(aq) + \_\_\_\_\_ + \_\_\_\_\_$$

d) chlorate(V) and chloride

$$ClO^{3-}(aq) + ?Cl^-(aq) + ?H^+(aq) \longrightarrow ?Cl_2(g) + \_\_\_\_\_$$

K3.3 Complete the balanced equations to show the reactions between the following pairs of substances in alkaline aqueous conditions (no fractions).

a) sulfite and bromate(V)

$$3\,SO_3^{2-}(aq) + \_\_\_\_\_ \longrightarrow 3\,SO_4^{2-}(aq) + \_\_\_\_\_$$

b) hydrogen peroxide and chromium(III)

$$3\,H_2O_2(aq) + ?Cr^{3+}(aq) + ?OH^-(aq) \longrightarrow ?CrO_4^{2-}(aq) + \_\_\_\_\_$$

c) chlorate(I) and iron(III)

$$?ClO^-(aq) + ?Fe(OH)_3(s) \longrightarrow ?FeO_4^{2-}(aq) + ?Cl_2(g) + \_\_\_\_\_ + \_\_\_\_\_$$

d) manganate(VI) and methanoate

$$?HCOO^-(aq) + MnO_4^{2-}(aq) \longrightarrow MnO_2(s) + OH^-(aq) + \_\_\_\_\_$$

10/12

K4 Disproportionation

K4.1 By assigning oxidation states to the relevant element in the following equations, show that disproportionation is occurring.

a) $Cl_2(aq) + H_2O(l) \longrightarrow HCl(aq) + HOCl(aq)$

b) $Cu_2SO_4(aq) \longrightarrow Cu(s) + CuSO_4(aq)$

c) $2\,CO(g) \longrightarrow C(s) + CO_2(g)$

d) $2\,HOF(aq) \longrightarrow H_2O(l) + OF_2(g)$

e) $5\,MnO_4^{2-}(aq) + 8\,H^+(aq) \longrightarrow Mn^{2+}(aq) + 4\,MnO_4^-(aq) + 4\,H_2O(l)$

K4.2 Complete and balance the following equations that represent disproportionation reactions.

a) $?H_2O_2(aq) \longrightarrow ?H_2O(l) +$ ____________

b) $?I_2(aq) + ?OH^-(aq) \longrightarrow$ ____________ $+ 5\,I^-(aq) + 3\,H_2O(l)$

c) $8\,S_2O_3^{2-}(aq) + 16\,H^+(aq) \longrightarrow S_8(s) +$ ____________ $+ 8\,H_2O(l)$

K4.3 For each of the disproportionation reactions below, write the separate oxidation and reduction half-equations occurring:

a) $3\,NO_2^-(aq) \longrightarrow 2\,NO(g) + NO_3^-(aq)$ in acidic conditions

i. Oxidation

ii. Reduction

b) $3\,IO_2^-(aq) \longrightarrow I^-(aq) + 2\,IO_3^-(aq)$ in alkaline conditions

i. Oxidation

ii. Reduction

Chapter L

Electrochemistry

L1 Electrode potential & cell potential

12/15

L1.1 Name the element whose reduction is used a standard by which all electrode potentials are measured.

L1.2 Give the concentration of hydrogen ions in a standard solution for reactions involving acid.

L1.3 The standard electrode potential, E°, for the reduction, $Br_2(aq) + 2e^- \longrightarrow 2Br^-(aq)$ is 1.09 V. Give the E° value for the reduction, $\frac{1}{2}Br_2(aq) + e^- \longrightarrow Br^-(aq)$.

L1.4 E° for the reaction, $Ce^{4+}(aq) + e^- \longrightarrow Ce^{3+}(aq)$ is 1.70 V. Give the E° value for the oxidation half-reaction, $Ce^{3+}(aq) \longrightarrow Ce^{4+}(aq) + e^-$.

Reaction	E°
$Zn^{2+}(aq) + 2e^- \longrightarrow Zn(s)$	−0.76 V
$Cr^{3+}(aq) + 3e^- \longrightarrow Cr(s)$	−0.74 V
$Fe^{2+}(aq) + 2e^- \longrightarrow Fe(s)$	−0.44 V
$Cu^{2+}(aq) + e^- \longrightarrow Cu^+(aq)$	+0.16 V
$Cu^{2+}(aq) + 2e^- \longrightarrow Cu(s)$	+0.34 V
$Cu^+(aq) + e^- \longrightarrow Cu(s)$	+0.52 V
$Fe^{3+}(aq) + e^- \longrightarrow Fe^{2+}(aq)$	+0.77 V
$Ag^+(aq) + e^- \longrightarrow Ag(s)$	+0.80 V
$Cr_2O_7^{2-}(aq) + 6e^- + 14H^+(aq) \longrightarrow 2Cr^{3+}(aq) + 7H_2O(l)$	+1.33 V

L1.5 Use the standard electrode potentials tabulated above to calculate the standard cell potentials due to the following reactions:

a) $Zn(s) + Cu^{2+}(aq) \longrightarrow Zn^{2+}(aq) + Cu(s)$

b) $Cu(s) + 2Ag^+(aq) \longrightarrow Cu^{2+}(aq) + 2Ag(s)$

c) $6Fe^{2+}(aq) + Cr_2O_7^{2-}(aq) + 14H^+(aq) \longrightarrow 6Fe^{3+}(aq) + 2Cr^{3+}(aq) + 7H_2O(l)$

d) $Fe^{2+}(aq) + Zn(s) \longrightarrow Fe(s) + Zn^{2+}(aq)$

L1.6 Using the data tabulated above, calculate the standard electrode potentials for the following reductions:

a) $\frac{1}{2}Zn^{2+}(aq) + e^- \longrightarrow \frac{1}{2}Zn(s)$

b) $Fe^{3+}(aq) + 3\,e^- \longrightarrow Fe(s)$

c) $Cu^{2+}(aq) + e^- \longrightarrow Cu^+(aq)$

d) $Cr_2O_7^{2-}(aq) + 14\,H^+(aq) + 12\,e^- \longrightarrow 2\,Cr(s) + 7\,H_2O(l)$

L1.7 Using the data tabulated above, calculate the standard cell potential for:

a) $2\,Cu^+(aq) \longrightarrow Cu(s) + Cu^{2+}(aq)$

b) $3\,Fe^{2+}(aq) \longrightarrow 2\,Fe^{3+}(aq) + Fe(s)$

c) $Ag^+(aq) + Cu^+(aq) \longrightarrow Ag(s) + Cu^{2+}(aq)$

$^{10}/_{12}$

L2 Free energy & K_c

L2.1 Use the standard electrode potentials tabulated in section L1 to find ΔG° for the following reductions:

a) $Ag^+(aq) + e^- \longrightarrow Ag(s)$

b) $Zn^{2+}(aq) + 2\,e^- \longrightarrow Zn(s)$

c) $Fe^{3+}(aq) + 3\,e^- \longrightarrow Fe(s)$

L2.2 Use the standard electrode potentials tabulated in section L1 to find ΔG° for the following reactions:

a) $Ag^+(aq) + Fe^{2+}(aq) \longrightarrow Fe^{3+}(aq) + Ag(s)$

b) $3\,Zn(s) + Cr_2O_7^{2-}(aq) + 14\,H^+(aq) \longrightarrow 3\,Zn^{2+}(aq) + 2\,Cr^{3+}(aq) + 7\,H_2O(l)$

c) $2\,Cr(s) + 3\,Cu^{2+}(aq) \longrightarrow 2\,Cr^{3+}(aq) + 3\,Cu(s)$

L2.3 Calculate E° for reactions with:

a) $\Delta G^\circ = -80\ kJ\,mol^{-1}$; $1\,e^-$ transferred

b) $\Delta G^\circ = 16\ kJ\,mol^{-1}$; $2\,e^-$ transferred

c) $\Delta G^\circ = -320\ J\,mol^{-1}$; $2\,e^-$ transferred

d) $K_c = 1.3 \times 10^6$ @ 298 K; $1\,e^-$ transferred

e) $K_c = 2.6 \times 10^{-6}$ @ 298 K; $3\,e^-$ transferred

f) $K_c = 120$ @ 400 K; $2\,e^-$ transferred

L3 Spontaneous redox reactions

5/7

1	As(V) → As(III)	+0.56	7	Cr(VI) → Cr(III)	+1.33
2	As(III) → As	+0.25	8	Cr(III) → Cr(II)	−0.41
3	As → As(−III)	−0.23	9	Cr(II) → Cr	−0.74
4	Mn(VII) → Mn(IV)	+1.70	10	I(VII) → I(V)	+1.60
5	Mn(IV) → Mn(II)	+1.23	11	I(V) → I	+1.19
6	Mn(II) → Mn	−1.19	12	I → I(−I)	+0.54

L3.1 Give the most powerful

a) Oxidizing agent in the table

b) Reducing agent in the table

L3.2 Select which of the states in the table could be oxidized spontaneously by elemental arsenic.

L3.3 Some reduction half-equations involving iron are shown below:

$FeO_4^{2-} + 6\,H^+ + 3\,e^- \rightarrow Fe^{3+} + 4\,H_2O$ $E^\theta = +2.20$ V

$Fe^{3+} + e^- \rightarrow Fe^{2+}$ $E^\theta = +0.77$ V

$Fe^{2+} + 2\,e^- \rightarrow Fe$ $E^\theta = -0.44$ V

a) Give the oxidation state of iron capable of oxidizing Mn(IV).

b) For each of the four elements listed, select the highest oxidation state that could be attained using Fe^{3+} as the oxidizing agent in aqueous acid: As, Cr, Mn, I

c) For each of the three elements listed, select the lowest oxidation state capable of oxidizing Fe^{2+} in aqueous acid: Cr, Mn, I

d) For each of the four elements listed, select the final oxidation state spontaneously reached via reduction of a higher state on exposure to excess metallic iron: As, Cr, Mn, I

Chapter M

Rate Laws, Graphs & Half-life

32/39

M1 Rate laws

		Rate law 1	rate = k
Equation 1	$A \longrightarrow B$	Rate law 2	rate = $k[A]$
Equation 2	$A + B \longrightarrow C$	Rate law 3	rate = $k[A]^2$
Equation 3	$A + B \longrightarrow C + D$	Rate law 4	rate = $k[A][B]$
Equation 4	$2A + B \longrightarrow C + D$	Rate law 5	rate = $k[A][B]^2$
		Rate law 6	rate = $k[A][B][cat]$

M1.1 a) A reaction described by equation 1 gets three times faster when the concentration of A is tripled. Give the order of reaction with respect to A, and the overall order of reaction.

b) If equation 2 proceeds as a single step, which rate law will it follow?

c) Which rate law(s) is/are second order overall?

d) In which rate law(s) is/are the units of the rate constant, k, $mol\,dm^{-3}\,s^{-1}$?

e) In rate law 6, the rate constant, k, has units which include dm raised to which power?

f) What is the order of reaction with respect to B in rate law 5?

g) Which rate law(s) describe a reaction in which reactant A always has constant half-life?

h) In rate law 2, if $[A] = 0.020\ mol\,dm^{-3}$, and the rate of reaction = $1.2 \times 10^{-3}\ mol\,dm^{-3}\,s^{-1}$, find the value of k.

i) In rate law 2, if k has a value of 150, find the rate of reaction when $[A] = 0.80\ mol\,dm^{-3}$.

j) In rate law 3, find [A] at which the reaction rate = 0.025 mol dm^{-3} s^{-1} if k = 0.0040 dm^3 mol^{-1} s^{-1}.

M1.2 Use the data in the table below to find the order of reaction with respect to R and S, the overall order of reaction, and the rate constant k. Include the correct unit in your answer for k.

[R]/mol dm^{-3}	[S]/mol dm^{-3}	Rate/mol dm^{-3} s^{-1}
0.010	0.010	2.4×10^{-4}
0.020	0.010	4.8×10^{-4}
0.010	0.025	6.0×10^{-4}

a) Order w.r.t. R: ______

b) Order w.r.t. S: ______

c) Overall order: ______

d) k: ______

M1.3 Use the data in the table below to find the order of reaction with respect to X and Y, the overall order of reaction, and the rate constant k. Include the correct unit in your answer for k.

[X]/mol dm^{-3}	[Y]/mol dm^{-3}	Rate/mol dm^{-3} s^{-1}
0.010	0.020	1.6×10^{-2}
0.020	0.020	6.4×10^{-2}
0.020	0.030	9.6×10^{-2}

a) Order w.r.t. X: ______

b) Order w.r.t. Y: ______

c) Overall order: ______

d) k: ______

M1.4 Use the data in the table below to find the order of reaction with respect to F and G, the overall order of reaction, and the rate constant k. Include the correct unit in your answer for k.

[F]/mol dm^{-3}	[G]/mol dm^{-3}	Rate/mol dm^{-3} s^{-1}
4.0×10^{-3}	1.0×10^{-2}	5.8×10^{-6}
2.0×10^{-3}	1.0×10^{-2}	2.9×10^{-6}
6.0×10^{-3}	2.0×10^{-2}	8.7×10^{-6}

a) Order w.r.t. F: ______

b) Order w.r.t. G: ______

c) Overall order: ______

d) k: ______

M1.5 Use the data in the table below to find the order of reaction with respect to A, B and the catalyst, X, the overall order of reaction, and the rate constant k. Include the correct unit in your answer for k.

[A]/mol dm^{-3}	[B]/mol dm^{-3}	[X]/mol dm^{-3}	Rate/mol dm^{-3} s^{-1}
0.50	0.080	0.0020	3.2×10^{-3}
0.50	0.080	0.0010	8.0×10^{-4}
0.75	0.080	0.0010	1.2×10^{-3}
0.75	0.040	0.0010	6.0×10^{-4}

a) Order w.r.t. A: ______

b) Order w.r.t. B: ______

c) Order w.r.t. X: ______

d) Overall order: ______

e) k: ______

M1.6 For a reaction following rate law 5, experimental data were collected. Fill in the gaps in the table below:

[A]/mol dm^{-3}	[B]/mol dm^{-3}	Rate/mol dm^{-3} s^{-1}
(a)	0.100	1.12×10^{-3}
0.25	0.100	(b)
0.25	0.050	1.40×10^{-4}
0.40	(c)	2.81×10^{-4}

M1.7 a) A reaction described by equation 1 shows the behaviour of graph (A) below. Which rate law does it follow?

b) A reaction described by equation 1 shows the behaviour of graph (B) below. Which rate law does it follow?

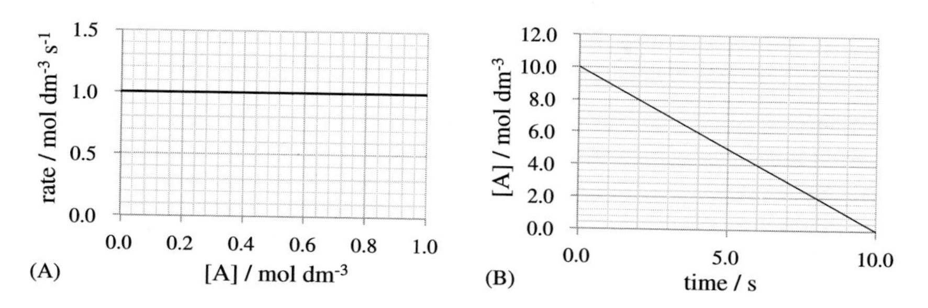

M1.8 A reaction described by equation 3 shows the behaviour of the graph below when a large excess of reactant B is used. Give the order of the reaction with respect to A.

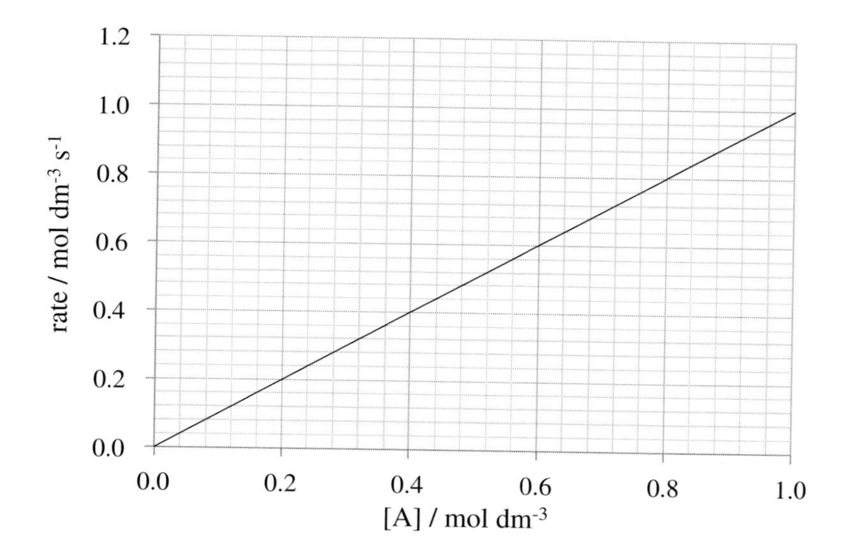

M1.9 A reaction described by equation 1 and following rate law 3 is investigated and the graph below is produced from the data obtained. Estimate the rate constant, k.

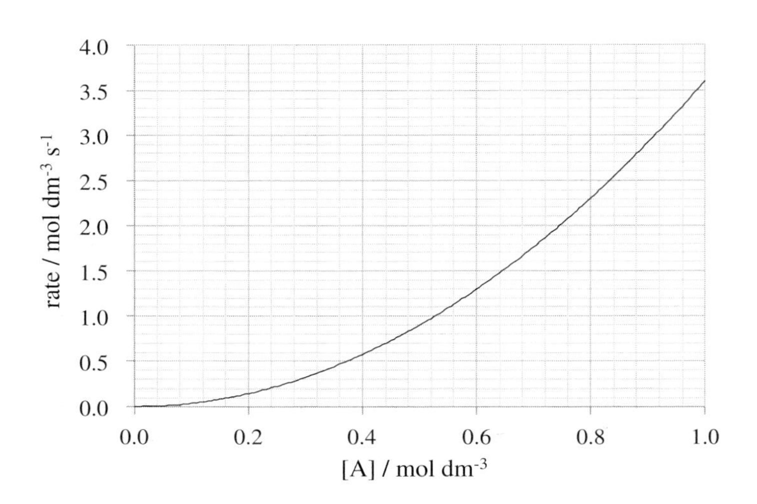

M1.10 A reaction described by equation 2 and obeying rate law 4 gave the following initial rates for different initial concentrations of A without varying the initial concentration of B. Estimate the initial concentration of B if the rate constant is 140 $dm^3\ mol^{-1}\ s^{-1}$.

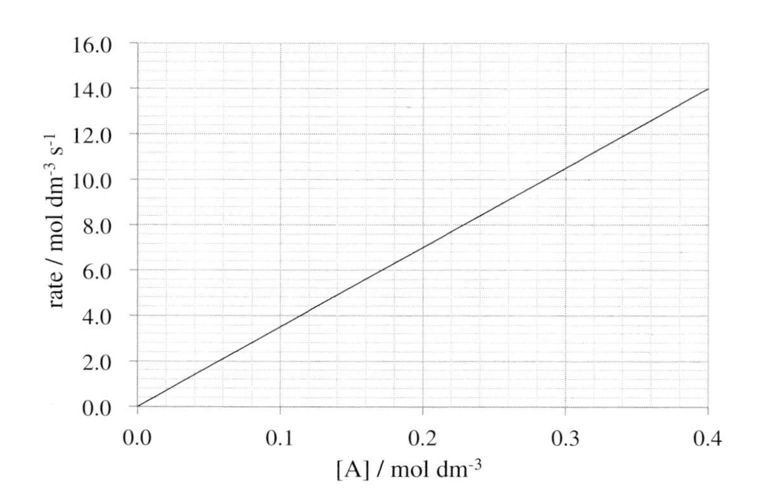

M2 Half-life

$^{18}/_{22}$

M2.1 a) A reactant with a constant half-life has a concentration of $0.60\ \text{mol dm}^{-3}$. Find its concentration after a period of 2 half-lives.

b) A first-order reaction has a reactant half-life of 25.0 minutes. If the initial concentration of reactant is $0.500\ \text{mol dm}^{-3}$, calculate the concentration of the reactant after 50.0 minutes.

c) A reactant with half-life 2.0 hours has an initial concentration of $6.4\ \text{g dm}^{-3}$. Calculate its concentration after 12 hours.

M2.2 a) A reactant has an initial concentration of $25.6\ \text{g dm}^{-3}$ and a half-life of 40.0 s. Calculate its concentration after 8.00 minutes.

b) A reactant with a half-life of 140 s has a concentration of $0.060\ \text{mol dm}^{-3}$ 7.0 minutes after the reaction began. Find the initial concentration of the reactant.

c) A reactant with a half-life of 18 minutes has a concentration of $1.3\ \text{g dm}^{-3}$ after 1.5 hours. If the reaction takes place in a constant volume of $400\ \text{cm}^3$ of solution, calculate the initial mass of reactant.

M2.3 a) In a first-order reaction, a reactant with initial concentration $0.80\ \text{mol dm}^{-3}$ is found after 3.0 hours of reaction to have a concentration of $0.050\ \text{mol dm}^{-3}$. Find the half-life of this reactant in minutes.

b) A reactant has a half-life of 30 s. If its initial concentration is $0.32\ \text{mol dm}^{-3}$, find the time taken from the beginning of the reaction for the reactant to reach a concentration of $2.5 \times 10^{-3}\ \text{mol dm}^{-3}$.

c) A first-order reaction is begun at 1400 h. If the half-life of a reactant is 75 minutes, find the time at which it will have been reduced to one eighth of its initial concentration.

d) A reaction occurs in a constant volume of $500\ \text{cm}^3$ of solvent. If a certain reactant has a half-life of 12 minutes, find the time taken, in minutes, to use up three quarters of that reactant.

e) A reactant with a constant half-life has an initial concentration of $0.30\ \text{mol dm}^{-3}$. If the reaction takes place in $200\ \text{cm}^3$ of solution, and there remains 15 mmol of reactant after 70 s, calculate its half-life.

M2.4 a) In a first-order reaction, it takes 48.0 s for a particular reactant to decrease in concentration from 2.00 mol dm^{-3} to 0.500 mol dm^{-3}. Calculate the decrease in concentration of this reactant over the next 48 s.

b) In a first-order reaction, it takes 40 s for a particular reactant to decrease in concentration from 1.20 mol dm^{-3} to 0.750 mol dm^{-3}. Calculate the concentration of this reactant after the next 80 s.

c) A reactant of initial concentration 0.400 mol dm^{-3} decreases in concentration by 0.150 mol dm^{-3} in 8.0 hours, obeying a first-order rate law. Calculate the further decrease in concentration over the next whole day.

M2.5 a) If a reactant with a constant half-life decreases in concentration by 0.100 mol dm^{-3} in 250 s, and by a further 0.080 mol dm^{-3} in the next 250 s, find its initial concentration.

b) A reactant in a first-order reaction has a concentration of 44.8 g dm^{-3} at the beginning of the reaction and a half-life of 15.0 s. If 0.35 g of reactant remains after 1.0 min, calculate the volume of solution in which the reaction is occurring.

M2.6 a) If the reaction, A(aq) + B(aq) $\longrightarrow$ C(aq) follows the rate law, rate = k[A], with an initial concentration of A of 0.280 mol dm^{-3}, which has a half-life of 0.940 h, find the concentration of C after 2.82 h, given that B is in excess and there was no C present at the start of the reaction.

b) The isomerisation of cyclopropane to propene occurs in the gas phase and follows a first-order rate law. If a mixture of the two gases has a mole fraction of 12% propene at 9:30 am and 89% propene at 7:00 pm on the same day, calculate the half-life of cyclopropane.

M2.7 A reaction described by the equation, 2 A(aq) $\longrightarrow$ 3 B(g) + 2 C(aq), follows the law, rate = k[A]. If the half-life of A is 23 s and the initial concentration of A is 0.75 mol dm^{-3}, find the volume of B produced at RTP by 1.8 dm^3 of solution in 46 s.

M2.8 The decomposition of hydrogen peroxide in aqueous solution is first order. The equation for the reaction is $H_2O_2(aq) \longrightarrow H_2O(l) + \frac{1}{2}O_2(g)$. If the half-life of hydrogen peroxide under certain conditions is 42 minutes, and 500 cm^3 of solution produces 6.20 dm^3 of oxygen at RTP in 1 hour and 24 minutes, find the initial concentration of hydrogen peroxide in mol dm^{-3}.

M2.9 Using the following graph, estimate the half-life of reactant A.

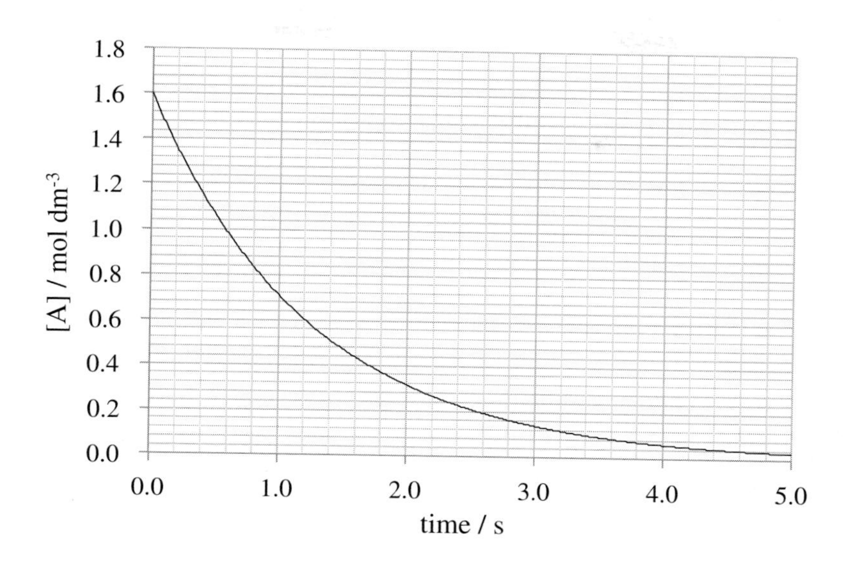

M2.10 Using the following graph, estimate the half-life of reactant A.

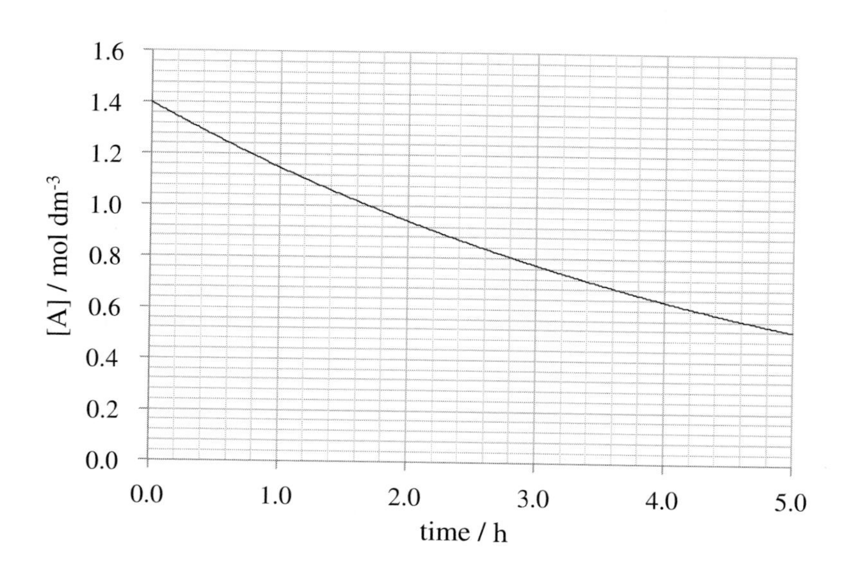

15/18

M3 The Arrhenius model

The Arrhenius equation is $k = Ae^{-E_A/RT}$. An Arrhenius plot is a graph of $\ln(k)$ against $1/T$ in K^{-1}.

M3.1 a) On a plot of $\ln(k)$ against $1/T$, what is the y-intercept?

b) Give the units of the gradient of an Arrhenius plot.

c) On a plot of $\ln(k)$ against $1/T$, what is the gradient?

M3.2 a) If the gradient of an Arrhenius plot is -1203 K, find the activation energy.

b) If the gradient of an Arrhenius plot is -4250 K, find the activation energy.

c) If a reaction has activation energy of 16.5 kJ mol^{-1}, find the expected gradient of an Arrhenius plot.

d) The pre-exponential factor, A, is found to have a value of 0.6 s^{-1} for a first-order reaction. Calculate the expected y-intercept of an Arrhenius plot.

M3.3 a) The y-intercept of an Arrhenius plot for a first-order reaction is at -2.30. Find the pre-exponential factor, A, according to the Arrhenius model.

b) The y-intercept of an Arrhenius plot for a second-order reaction is at 3.20. Find the pre-exponential factor, A, according to the Arrhenius model.

M3.4 a) The rate constant, k, for a first-order reaction is found to be 26.0 s^{-1} at 290 K. If the pre-exponential factor is 0.025 s^{-1}, find the activation energy.

b) The rate constant, k, for a second-order reaction is found to be 0.050 dm^3 mol^{-1} s^{-1} at 300 K. If the activation energy is 2.50 kJ mol^{-1}, find the value of the pre-exponential factor, A, in dm^3 mol^{-1} s^{-1}.

M3.5 A first order reaction has pre-exponential factor 8.0 s^{-1} and activation energy 4.8 kJ mol^{-1}. Find the rate constant at:

a) 290 K

b) 900 K

M3.6 If a reaction has activation energy 14.0 kJ mol^{-1}, and a pre-exponential factor of 120 s^{-1}, find the temperature at which the rate constant is equal to 2.00 s^{-1}.

M3.7 A reaction is found to have a rate constant of 1.25×10^{-3} dm^6 mol^{-2} s^{-1} at 400 K and 1.60×10^{-3} dm^6 mol^{-2} s^{-1} at 500 K. Find the activation energy. Find the pre-exponential factor, A. Give the overall order of reaction.

M3.8 Using the following graph, find the activation energy and the value of the pre-exponential factor for the reaction described by it.

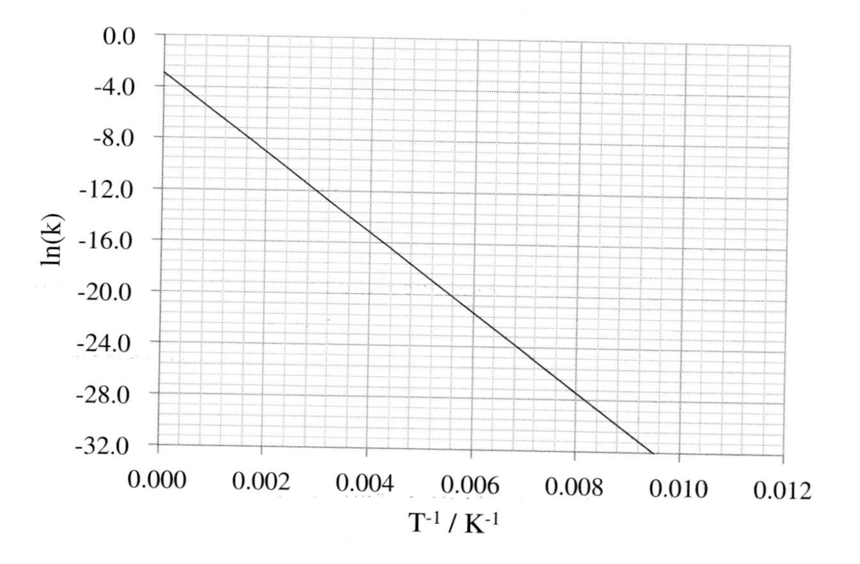

M3.9 Using the following graph, find the activation energy and the value of the pre-exponential factor for the reaction described by it.

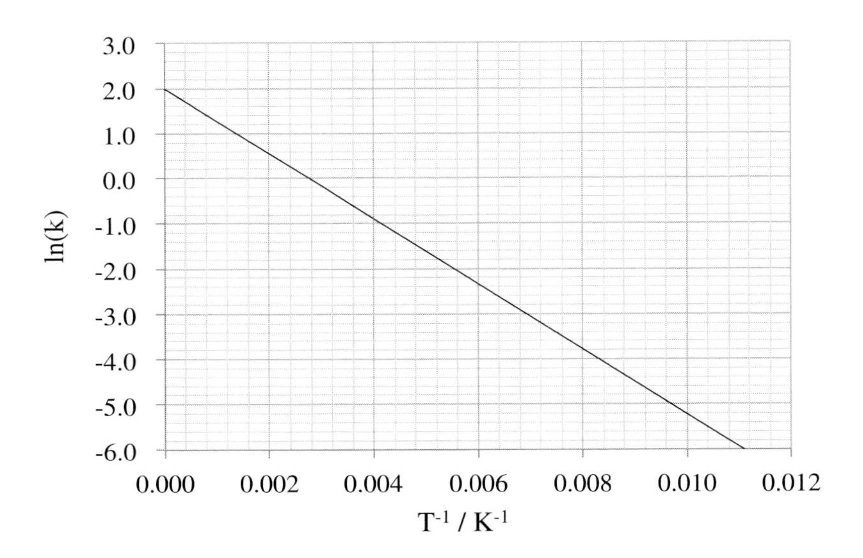

M3.10 Using the following graph, find the activation energy and the value of the pre-exponential factor for the reaction described by it.

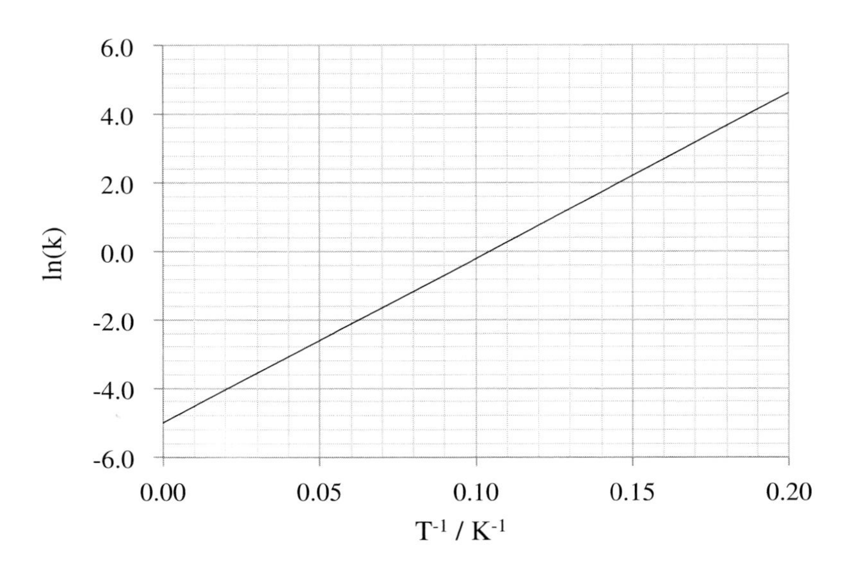

M4 Catalysis

16/19

M4.1 The iodination of propanone, $C_3H_6O + I_2 \longrightarrow C_3H_5OI + HI$, when catalysed in aqueous conditions, obeys the rate law, rate $= k[C_3H_6O][HCl]$.

a) Identify the catalyst in this reaction.

b) Is the catalyst homogeneous or heterogeneous?

c) If the catalyst has an initial concentration of 0.020 mol dm^{-3}, give the concentration of the catalyst when the concentration of propanone has decreased to one quarter of its original value.

M4.2 A reversible reaction reaches equilibrium when the forward and backward rates are both equal to 0.26 mol dm^{-3} s^{-1}. If the addition of a heterogeneous catalyst changes the rate law from second-order to zero-order, and increases the forward rate by a factor of 200, calculate the backward rate of the catalysed reaction.

M4.3 Vanadium(V) oxide is used as a catalyst in the oxidation of sulfur(IV) oxide to sulfur(VI) oxide. Complete the table with 'increases', 'decreases' or 'stays the same' to show the effects of adding the catalyst.

Yield at equilibrium	(a)
Time taken to reach equilibrium	(b)
Rate of reaction	(c)
Sulfur(VI) oxide produced per kg of sulfur(IV) oxide used	(d)
Amount of sulfur(VI) oxide produced per day	(e)
Mass of vanadium(V) oxide present	(f)

M4.4 Choose the correct continuations. A catalyst increases the rate of a reaction...

A. ...without taking part in the reaction.

B. ...without being used up.

C. ...by shifting equilibrium towards the products.

D. . . by lowering the activation energy when reactants collide.

E. . . by providing an alternative reaction route of lower activation energy.

F. . . by increasing the energy of the reactant particles.

G. . . by raising the energy of the transition state.

H. . . through the formation of an intermediate.

M4.5 A set of pupils drew up this list of what they thought might be potential catalysts:

1) Heat
2) Manganese(IV) oxide
3) Iron
4) Nickel
5) Vanadium(V) oxide
6) Catalase
7) Light
8) Gold
9) Chlorine atoms

Fill in the seven spaces in the table below by selecting from the numbers above:

Catalysed reactions	Number(s) from list
$2\,H_2O_2 \longrightarrow 2\,H_2O + O_2$	(a)
$C_{15}H_{29}COOH + H_2 \longrightarrow C_{15}H_{31}COOH$	(b)
$N_2 + 3\,H_2 \longrightarrow 2\,NH_3$	(c)
$2\,SO_2 + O_2 \longrightarrow 2\,SO_3$	(d)
$2\,O_3 \longrightarrow 3\,O_2$	(e)
None of the above, but could be a catalyst for another reaction.	(f)
Could never be a catalyst.	(g)

M4.6 Complete the following description of heterogeneous catalysis using numbers relating to the list shown below:
A heterogeneous catalyst works best when it has a large _____ so that

many of its ______ are exposed to the reactants. In the first stage, reactants are ______ onto the catalytic surface, weakening internal bonds in the reactant particles. In the second stage, the particles react to form products. The ______ of the slowest step in this reaction is lower than that of the slowest step in the uncatalysed reaction. The third stage is the ______, or release, of product particles from the surface. This ______ the surface ready for further reactions. If the products are not released, or if some contaminant binds to the surface, further catalytic activity is impeded and the catalyst has been ______.

1) mass	6) destruction	11) active sites
2) desorption	7) poisoned	12) volume
3) enthalpy change	8) absorbed	13) adsorbed
4) rate	9) regenerates	14) activation energy
5) density	10) surface area	15) corroded

M5 Michaelis-Menten kinetics

15/18

In the Michaelis-Menten model of enzyme kinetics:

$$-\frac{d[S]}{dt} = \frac{k_1[E][S]}{K_M + [S]}$$

where E and S are enzyme and substrate, and K_M is the Michaelis constant.

M5.1 a) If K_M is small compared to [S], give the approximate overall order of the rate law.

b) If K_M is small compared to [S], give the order of reaction with respect to S.

M5.2 If K_M is large compared to [S], give the approximate overall order of the rate law.

M5.3 If [E] remains constant and [S] has a constant half-life, is K_M small or large compared to [S]?

M5.4 When [S] is equal to K_M, the rate of reaction = nk_1[E]. Give the value of n.

M5.5 Complete the following table by filling in the missing value in each row.

Rate / $\mathrm{mol\,dm^{-3}\,s^{-1}}$	**k_1 / $\mathrm{s^{-1}}$**	**K_M / $\mathrm{mol\,dm^{-3}}$**	**[E] / $\mathrm{mol\,dm^{-3}}$**	**[S] / $\mathrm{mol\,dm^{-3}}$**
(a)	1.50×10^{15}	0.0120	1.60×10^{-8}	2.50×10^{-3}
(b)	8.2×10^{14}	6.0×10^{-3}	1.4×10^{-9}	1.2×10^{-2}
1.76×10^{7}	(c)	3.20×10^{-5}	1.10×10^{-10}	4.00×10^{-2}
1.60×10^{3}	4.40×10^{12}	0.160	(d)	1.90×10^{-3}
1.20×10^{6}	(e)	2.00×10^{-7}	1.20×10^{-9}	0.0110
8.40	2.10×10^{11}	(f)	4.00×10^{-10}	2.00×10^{-5}
47.9	1.30×10^{14}	2.20	5.40×10^{-9}	(g)
6.37	3.10×10^{8}	(h)	2.20×10^{-8}	6.00×10^{-3}

M5.6 Give the gradient and intercept of a graph of $\frac{1}{\text{rate}}$ against $\frac{1}{[\mathrm{S}]}$ that would be obtained using the reaction given in the last row of the table above.

M5.7 A graph of $\frac{1}{\text{rate}}$ against $\frac{1}{[\mathrm{S}]}$ has a gradient of 4.2×10^{-7} s. If $k_1 = 1.2 \times 10^{13}\ \mathrm{s^{-1}}$ and $[\mathrm{E}] = 8.0 \times 10^{-8}\ \mathrm{mol\,dm^{-3}}$, find K_M.

M5.8 A graph of $\frac{1}{\text{rate}}$ against $\frac{1}{[\mathrm{S}]}$ has an intercept of $1.2 \times 10^{-3}\ \mathrm{dm^3\,s\,mol^{-1}}$ and a gradient of 6.0×10^{-6} s. If $[\mathrm{E}] = 4.0 \times 10^{-9}\ \mathrm{mol\,dm^{-3}}$, find k_1 and K_M.

M5.9 A graph of $\frac{1}{\text{rate}}$ against $\frac{1}{[\mathrm{S}]}$ has an intercept of $3.2 \times 10^{-4}\ \mathrm{dm^3\,s\,mol^{-1}}$. If $k_1 = 6.0 \times 10^{12}\ \mathrm{s^{-1}}$, find [E].

M5.10 Use the following graph to estimate K_M. Given that $k_1 = 1.2 \times 10^{14}$ s^{-1}, estimate [E].

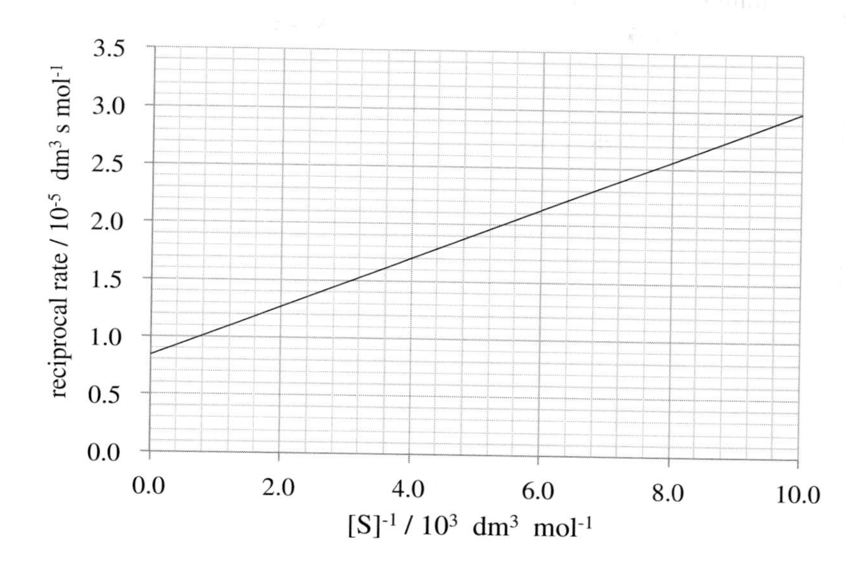